W9-ADS-769

Lecture Notes in Mathematics

Edited by A. Dold and B. Eckmann

469

Ernst Binz

Continuous Convergence on C(X)

Springer-Verlag
Berlin · Heidelberg · New York 1975

Author
Prof. Dr. Ernst Binz
Universität Mannheim (WH)
Lehrstuhl für Mathematik I
68 Mannheim
Schloß
BRD

Library of Congress Cataloging in Publication Data

Binz, Ernst, 1939-
Continuous convergence on C(X)

(Lecture notes in mathematics ; 469)
Bibliography: p.
Includes index.
1. Function spaces. 2. Convergence.
3. Topological algebras. I. Title. II. Series:
Lecture notes in mathematics (Berlin) ; 469.
QA3.L28 no. 469 [QA323] 510'8s [512'.55] 75-16495

AMS Subject Classifications (1970): 22-02, 22A99, 46-02, 46E99, 46H99, 46M99, 54-02, 54A20, 54C35, 54C99, 54H10

ISBN 3-540-07179-2 Springer-Verlag Berlin · Heidelberg · New York
ISBN 0-387-07179-2 Springer-Verlag New York · Heidelberg · Berlin

This work is subject to copyright. All rights are reserved, whether the whole or part of the material is concerned, specifically those of translation, reprinting, re-use of illustrations, broadcasting, reproduction by photocopying machine or similar means, and storage in data banks.

Under § 54 of the German Copyright Law where copies are made for other than private use, a fee is payable to the publisher, the amount of the fee to be determined by agreement with the publisher.

© by Springer-Verlag Berlin · Heidelberg 1975

Printed in Germany

Offsetdruck: Julius Beltz, Hemsbach/Bergstr.

MATH.-SCI.
1413892
QA
323
.B55

T o :

Erika, Barbara and Dieter

ACKNOWLEDGEMENT

Much of the material presented in these notes was worked out in collaboration with K.Kutzler and my former students: H.P.Butzmann, W.A.Feldman and M.Schroder. I am greatly indebted to all of them.

Special thanks go to H.P.Butzman and B.Müller for their criticisms, suggestions and the painful job of proofreading.

The suggestion of C.H.Cook to write these notes in English provided him and Dany Gulick with some work of linguistic adjustments. I am grateful to both of them.

The manuscript was typed by Mrs.K.Bischoff. I would like to thank her very much for the care she took with this job.

Math
Sep

TABLE OF CONTENTS

INTRODUCTION

Some parts of functional analysis and general topology are devoted to the relationship between a completely regular topological space X and $C_{co}(X)$. By $C_{co}(X)$, we mean the $\mathbb{R}$-algebra $C(X)$ of all continuous, real-valued functions of X, equipped with the topology of uniform convergence on compact subsets of X. Many of these investigations of this relationship are hindered by the fact that the evaluation map

$$\omega : C_{co}(X) \times X \longrightarrow \mathbb{R}$$

(sending each pair $(f,p) \in C(X) \times X$ into $f(p)$) is not continuous. Another handicap is that, in general, $C_{co}(X)$ is not complete. In these notes we replace the concept of uniform convergence on compact subsets of X by the concept of continuous convergence. This type of convergence on $C(X)$ does not arise from a topology. However, it is generated by a so-called convergence structure, a notion which generalizes that of a topology. The convergence structure of continuous convergence (the continuous convergence structure) is finer than the topology of compact convergence and coincides with it when X is a locally compact topological space. The algebra $C(X)$, endowed with the continuous convergence structure, yields a complete convergence algebra, denoted by $C_c(X)$, and a continuous evaluation map

$$\omega : C_c(X) \times X \longrightarrow \mathbb{R}.$$

The convergence algebra $C_c(X)$ carries the coarsest among all the convergence structures Λ on $C(X)$ for which $\omega : C_\Lambda(X) \times X \longrightarrow \mathbb{R}$ is continuous. This fact provides $C_c(X)$ with many convenient properties. However, difficulties occur in many approximation problems especially in the context of a Stone-Weierstrass type of theorem.

The purpose of these notes is to present the foundations of the interactions between a general convergence space X and $C_c(X)$.
In Chapter o, we introduce the theory of convergence spaces (convergence spaces, continuous convergence structure, etc.). In Chapter 1, we collect some properties of $C(X)$ needed in the subsequent chapters (e.g. Stone-Čech- and realcompactifications etc.)
We demonstrate in Chapter 2, that, for a completely regular topological space X, there is (in general) no $\mathbb{R}$-vector space topology T on $C(X)$ for which

$$\omega : C_T(X) \times X \longrightarrow \mathbb{R}$$

is continuous.
In the third chapter, we exhibit and study a special class of convergence spaces, the class of c-embedded spaces. For any two such spaces X and Y, the convergence algebras $C_c(X)$ and $C_c(Y)$ are bicontinuously isomorphic iff X and Y are homeomorphic. This class turns out to be very large: Any $C_c(Z)$, where Z is an arbitrary convergence space, can be represented by $C_c(Z')$, where Z' is a c-embedded space. Among other topological spaces any completely regular topological space is c-embedded. The structure of c-embedded spaces is expressed by Schroder's theorem, which asserts that any c-embedded space is the projective limit of inductive limits of compact topological spaces. To find these topological results, we have to develop the functional analytic apparatus of $C_c(X)$. Along the way, we observe that the relationship between a c-embedded space X and $C_c(X)$ is an extension of the classical correspondence between a compact topological space Y and the Banach algebra $C_c(Y)$.
The problem of studying which convergence $\mathbb{R}$-algebras are of the form $C_c(Y)$ is the principal intent of Chapter 4. However, we restrict ourselves mostly to subalgebras of $C_c(X)$. These investigations have, of

course, some similarities to the Gelfand theory. We conclude the chapter with a study of the c-embeddedness of general function spaces. There the reader may notice that the category of c-embedded spaces is cartesian closed.

Chapter 5 is devoted to parts of the dictionary of topological properties of c-embedded spaces X and functional analytic properties of $C_c(X)$. In this context we will in particular characterize normal, separable metric and Lindelöf spaces. Necessary and sufficient conditions are given on a completely regular topological space X in order that $C_c(X)$ be representable as an inductive limit of topological $\mathbb{R}$-vector spaces.

In the appendix, we develop some results on the linear and Pontryagin duality of $C_c(X)$ and of general topological $\mathbb{R}$-vector spaces. Here the respective dual spaces will be endowed with the continuous convergence structure.

Finally we remark that the items in the Bibliography, not quoted in the text, contain interesting suplements to the material presented in these notes.

0. CONVERGENCE SPACES

In this introductory chapter we collect some elementary material on convergence spaces.

0.1 Convergence spaces

Roughly speaking, a convergence space is a set, together with a concept of convergence. This concept is based on the notion of a filter, in the sense of [Bou], to whom we refer for terminology and general properties of filters.

We denote by $\mathcal{F}(X)$ the set of all filters on X, and by $\mathcal{P}(\mathcal{F}(X))$ the power set of $\mathcal{F}(X)$. The definition of a convergence structure and a convergence space now reads as follows:

Let X be a set. A map $\Lambda : X \longrightarrow \mathcal{P}(\mathcal{F}(X))$ is said to be a <u>convergence structure</u> if the following properties hold for any point $p \in X$:

(i) $\dot{p} \in \Lambda(p)$.

(ii) $\Phi, \Psi \in \Lambda(p) \Longrightarrow \Phi \wedge \Psi \in \Lambda(p)$.

(iii) $\Phi \in \Lambda(p)$ and $\Psi \in \quad (X)$ with $\Psi \geqslant \Phi \Longrightarrow \Psi \in \Lambda(p)$.

Here $\Phi \wedge \Psi$ denotes the infimum of Φ and Ψ and $\dot{p}$ stands for the filter formed by all supersets of $\{ p \} \subset X$.

The pair (X,Λ) is named a <u>convergence space</u>. The filters in $\Lambda(p)$ are said to be <u>convergent</u> to p.

We usually write $\Phi \longrightarrow p$ instead of $\Phi \in \Lambda(p)$.

To illustrate the terms introduced, we list the following two examples of convergence structures:

1) Let X be a (non-empty) topological space. By assigning to each

point $p \in X$ the set $\mathcal{T}(p)$ of all filters on X which converge with respect to the given topology to the point p, we obtain a convergence structure. Hence any topology can be interpreted as a convergence structure. We therefore call a convergence structure <u>topological</u> or simply a <u>topology</u> if the convergent filters are precisely those of a topology.

2) To construct a convergence structure which is not topological, we assume that E is an infinite dimensional vector space over $\mathbb{R}$ (the reals). For the filters convergent to $p \in E$ we choose the filters on E having a basis $\mathcal{B}$, where $\mathcal{B}$ is a filter on some finite dimensional subspace R of E which converges on R to the point p with respect to the natural topology. It is an easy exercise to check that the convergence structure defined in this way is not a topology.

More examples of convergence structures will appear throughout the text, others and applications of them can be found in [Ba], [Ke], [Ma], and [Wl], as well as in other items quoted in the "Bibliography".

To reach our goal, we need to develop the general theory of convergence structures a little further. In doing so we follow closely the pattern of general topology laid out in [Bou].

To simplify the notation, we normally use the symbol X to denote the convergence space (X,Λ).

Let F be a subset of X. The point $p \in X$ is a <u>point of adherence</u> of F if it admits a filter converging to p which has a trace on F. The adherence of F, the set of all points adherent to F, is denoted by $\bar{F}$.

As in general topology, we call a subset A of X <u>closed</u> if $A = \bar{A}$. The complements of the closed subsets are said to be <u>open</u> sets. Those are characterized by saying that a set $O \subset X$ is open if it belongs to every filter which converges to a point of O.

The collection of open sets of a convergence space X fulfills the axioms of a topology. This topology is called the topology associated to the convergence structure of X. For the resulting topological space we use the symbol X_T and refer to it as the topological space associated to X.

A convergence space X is Hausdorff if a filter on X converges to at most one point.

A map f from a convergence space X into another such space Y is continuous at a point $p \in X$ if for any filter Φ convergent to p the filter $f(\Phi)$ (generated by all sets $f(F)$, where F runs through Φ) converges to $f(p)$. If f is continuous at every point $p \in X$, then f is said to be continuous. If f is bijective and both f and f^{-1} are continuous, then we speak of f as a homeomorphism.

As a simple exercise one may prove that a map f defined on a convergence space X with values in a topological space Y is continuous iff $f : X_T \longrightarrow Y$ is continuous.

If for two convergence structures Λ_1 and Λ_2 on a set Z the identity

$$id : (Z,\Lambda_1) \longrightarrow (Z,\Lambda_2)$$

is continuous, then Λ_1 is said to be finer then Λ_2, expressed in symbols by $\Lambda_1 \geqslant \Lambda_2$ or $\Lambda_2 \leqslant \Lambda_1$.

Next we introduce the notions of initial and final convergence structures and some concepts based on them. Let

$$f_\iota : X \longrightarrow Y_\iota$$

be a family of maps from a set X into a family $\{ Y_\iota \}_{\iota \in I}$ of convergence spaces. To any point $p \in X$ we assign all those filters Φ on X for which $f_\iota(\Phi)$ converges to $f_\iota(p)$ for every $\iota \in I$. The convergence structure Ω on X defined in this way is the initial convergence structure induced by the family $\{ f_\iota \}_{\iota \in I}$. It is of course the coarsest of all the convergence structures on X which allow every f_ι to be continuous.

Hence a map f from a convergence space S into (X,Ω) is continuous iff $f_\iota \circ f$ is continuous for every $\iota \in I$. This universal property is characteristic for the initial convergence structure.

Clearly if every convergence space of the family $\{ Y_\iota \}_{\iota \in I}$ is topological then Ω is a topology.

Based on the notion of the initial convergence structure one defines subspaces and products in the obvious way. A subset F of a convergence space X is turned into a <u>subspace</u> of X if it is endowed with the initial convergence structure induced by the inclusion map.

The <u>product</u> $\prod_{\iota \in I} X_\iota$ of a family $\{ X_\iota \}_{\iota \in I}$ of convergence spaces is the product of the underlying sets of the family endowed with the initial convergence structure induced by the family of all the canonical projections.

For a family $\{ X_\iota \}_{\iota \in I}$ of convergence spaces and a family of maps $\{ f_\iota \}_{\iota \in I}$ into a set Y, we define the <u>final convergence structure</u> Ω' induced by the family $(f_\iota)_{\iota \in I}$ by stating: A filter Φ on Y converges to $p \in Y$ iff $\Phi \geqslant \dot{p}$ or $\Phi \geqslant \bigwedge_{i=1}^{n} f_{\iota_i}(\Psi_i)$, where the filters Ψ_i converge to a preimage under f_{ι_i} of p for $\iota_i \in I$ and $i = 1,\ldots,n$. The universal property characterizing the final convergence structure reads as follows:

A map f from (Y,Ω') into a convergence space Z is continuous iff $f \circ f_\iota$ is continuous for every $\iota \in I$.

The <u>quotient convergence structure</u> on the set X' of all equivalence classes in a convergence space X is the final structure induced by the canonical projection.

For a detailed study of the general theory of convergence spaces we refer to [C,F], [Bi,Ke], [Fi], [Ko], and [Wo].

0.2 The structure of continuous convergence

The <u>structure</u> of <u>continuous convergence</u> Λ_c will be defined on the function space $C(X,Y)$, the set of all continuous maps from a convergence space X into a convergence space Y. It will be a convergence structure closely connected with the <u>evaluation map</u>

$$\omega : C(X,Y) \times X \longrightarrow Y,$$

which sends each couple (f,p) into $f(p)$.

In fact, a filter Θ belongs to $\Lambda_c(f)$ if

$$\omega(\Theta \times \Phi) \longrightarrow f(p) \in Y$$

for any filter Φ converging to $p \in X$ and any point $p \in X$. The symbol $\Theta \times \Phi$ denotes the filter generated by all sets of the form $T \times F$, where $T \in \Theta$ and $F \in \Phi$. The set $C(X,Y)$ equipped with Λ_c is denoted by $C_c(X,Y)$. If Y is Hausdorff clearly Λ_c is Hausdorff too. We will reach the converse below.

A convergence structure Λ on $C(X,Y)$ is called <u>ω-admissible</u> if

$$\omega : (C(X,Y),\Lambda) \times X \longrightarrow Y$$

is continuous (the product carries the product convergence structure).

It is a simple exercise to verify that Λ_c is ω-admissible and moreover Λ_c is coarser than every ω-admissible convergence structure on $C(X,Y)$.

A subset $H \subset C(X,Y)$ regarded as a subspace of $C_c(X,Y)$ is denoted by H_c and said to be equipped with the structure of continuous convergence.

A characteristic universal property of H_c is the following one: A map g from a convergence space S into H_c is continuous iff the composition $\omega \circ (g \times id_X)$ of the map

$$g \times id_X : S \times X \longrightarrow H_c \times X$$

and the restriction of ω onto H_c, again denoted by

$$\omega : H_c \times X \longrightarrow Y ,$$

is continuous.

Using this universal property one easily proves that the set K of all constant maps endowed with the structure of continuous convergence is, via the canonical map from K into Y, homeomorphic to Y. Hence $C_c(X,Y)$ is Hausdorff iff Y is Hausdorff.

Let us point out here that in case X is a locally compact topological space and Y coincides with $\mathbb{R}$, the convergence structure Λ_c on $C(X,\mathbb{R})$ is the topology of compact convergence [Schae].

Detailed studies of the structure of continuous convergence ("Limitierung der stetigen Konvergenz") can be found in [Ba], [C,F], [Bi,Ke], and [Po].

1. FUNCTION ALGEBRAS

As indicated in the preface, one of our major aims is to describe the relationship between a convergence space X and $C_c(X)$ (we use C(X) as a shorthand of $C(X,\mathbb{R})$). To do so, we firstly have to turn our attention to the relationship between X and C(X). This will be the purpose of the following two sections.

1.1 The completely regular topological space associated to a convergence space

The set C(X) endowed with the pointwise defined operations is a lattice $\mathbb{R}$-algebra [G,J].

We allow ourselves to use the term "function algebra of X" or on some occasions simply "function algebra" to speak of an $\mathbb{R}$-algebra of the form C(X).

To get a close relationship between X and C(X) one should at least know that C(X) is able to separate points. This means that $f(p)=f(q)$ for each $f \in C(X)$ should imply $p=q$. However, it is easy to construct examples of even topological spaces for which their function algebras do not separate points [G,J]. To study C(X) we will therefore associate to X a certain quotient space X_s which has a point-separating function algebra isomorphic to C(X). Moreover, the topology on X_s will be determined by C(X).

In the set X we define an equivalence relation by saying, that any two points p and q are equivalent iff $f(p)=f(q)$ for all $f \in C(X)$.

We would have obtained the same equivalence classes if we had required that any two points p and q are equivalent if $f(p)=f(q)$ for each $f \in C^o(X)$ where $C^o(X)$ denotes the lattice subalgebra of C(X) consisting of all bounded elements of C(X).

Let us denote the set of all equivalence classes by $\bar{X}$ and the canonical projection from X onto $\bar{X}$ by π.

Each function $f \in C(X)$ induces a function

$$\bar{f} : \bar{X} \longrightarrow \mathbb{R},$$

by defining $\bar{f}(\bar{p})=f(p)$ for each $\bar{p} \in \bar{X}$ (by $\bar{p}$ we denote the equivalence class of a point $p \in X$).

The set $\bar{X}$ together with the initial convergence structure induced by the family $\{ \bar{f} \}_{f \in C(X)}$ is a completely regular topological space denoted by X_s. The reader may verify that the initial convergence structures on X induced by $\{ \bar{f} \}_{f \in C(X)}$ and $\{ \bar{f} \}_{f \in C^o(X)}$ are identical.

A topological space Z is said to be <u>completely regular</u> if each set consisting of one point is closed and moreover, if any point p and any closed set F not containing p can be strongly separated by some $f \in C(Z)$. By strongly separated we mean $f(p)=1$ and $f(q)=o$ for each $q \in F$. Evidently, X_s is completely regular. It is called the <u>completely regular topological space associated</u> to X.

If $\pi : X \longrightarrow X_s$ is an injection it need not to be a homeomorphism as the following example (in [G,J], p. 50) shows:

Let S denote the subspace of $\mathbb{R} \times \mathbb{R}$ obtained by deleting (o,o) and all $(\frac{1}{n},q)$ where n runs through the natural numbers $\mathbb{N}$ and q through $\mathbb{R}$. Define $g : S \longrightarrow \mathbb{R}$ by $g(p,q)=p$ for each pair $(p,q) \in \mathbb{R} \times \mathbb{R}$. Endow the set of the reals with the finest topology for which g is continuous. Call the resulting topological space E. As one easily verifies E is Hausdorff and has the same continuous real-valued functions as $\mathbb{R}$. Thus $E_s = \mathbb{R}$. The set $\{ \frac{1}{n} \mid n \in \mathbb{N} \}$ is closed in E. But it can not be strongly separated from o. Thus E is not completely regular. This example shows in addition, that the quotient of a completely regular topological space is in general not completely regular (the quotient is taken within the category of topological spaces).

Clearly if X is a completely regular topological space, then $\pi : X \longrightarrow X_s$ is a homeomorphism.

The $\mathbb{R}$-algebra homomorphism (sending unity into unity)

$$\pi^* : C(X_s) \longrightarrow C(X),$$

defined by $\pi^*(g) = g \circ \pi$ for each $g \in C(X_s)$, is an $\mathbb{R}$-algebra isomorphism.

To simplify terminology let us replace "$\mathbb{R}$-algebra homomorphism sending unity into unity" just by "<u>homomorphism</u>".

Now let X and Y be convergence spaces and let

$$f : X \longrightarrow Y$$

be a continuous map. By

$$f^* : C(Y) \longrightarrow C(X)$$

we mean the homomorphism defined by $f^*(g)=g \circ f$ each $g \in C(Y)$. Clearly f^* sends bounded functions into bounded functions. Instead of using $f^* | C^o(Y)$ we often will write f^* only.

We again leave it to the reader to check that if X and Y are completely regular topological spaces, then f is a homeomorphism iff either

$$f^* : C(Y) \longrightarrow C(X)$$

or

$$f^* : C^o(Y) \longrightarrow C^o(X)$$

is an isomorphism.

Is it true that any homomorphism $h : C(Y) \longrightarrow C(X)$ (resp. $h : C^o(Y) \longrightarrow C^o(X)$) is of the form f^* for some map $f : X \longrightarrow Y$, where X and Y are any two completely regular topological spaces? If it were true, then any two completely regular topological spaces would be homeomorphic if their function algebras are isomorphic. The next section gives the answer.

1.2 Real-compactification and Stone-Čech compactification of a completely regular topological space

To investigate the problems connected with the question asked above, we exhibit spaces depending only on $C(X)$ and $C^o(X)$, respectively .

The underlying sets of the spaces we are looking for are $\text{Hom}\, C(X)$ and $\text{Hom}\, C^o(X)$, the collection of all real-valued homomorphisms of $C(X)$ and $C^o(X)$, respectively.

The map

$$j^* : \text{Hom}\, C(X) \longrightarrow \text{Hom}\, C^o(X),$$

assigning to each element $h \in \text{Hom}\, C(X)$ its restriction to $C^o(X)$, is injective. Indeed, let $h_1, h_2 \in \text{Hom}\, C(X)$ be distinct elements and let $f \in C(X)$ be such that

$$h_1(f) = o \quad \text{and} \quad h_2(f) > o.$$

The function

$$((-h_2(f) \cdot \underline{1}) \vee f) \wedge h_2(f) \cdot \underline{1},$$

where $\underline{1}$ denotes the constant function on X assigning the value 1, is certainly bounded and continuous. Since every homomorphism on $C(X)$ is a lattice homomorphism ([G,J], theorem 1.6), we have

$$h_1(g) = o \quad \text{and} \quad h_2(g) = h_2(f) > o.$$

From now on let us identify each $h \in \text{Hom}\, C(X)$ with $j^*(h)$. We equip both sets $\text{Hom}\, C(X)$ and $\text{Hom}\, C^o(X)$ with the following topologies. Every function $f \in C(X)$ induces a function

$$d'(f) : \text{Hom}\, C(X) \longrightarrow \mathbb{R},$$

sending each $h \in \text{Hom}\, C(X)$ into $h(f)$. Similarly we define

$$d'(f) : \text{Hom}\, C^o(X) \longrightarrow \mathbb{R}$$

if $f \in C^o(X)$.

Let us denote by $\mathrm{Hom}_s C(X)$ and $\mathrm{Hom}_s C^o(X)$ the sets $\mathrm{Hom}\, C(X)$ and $\mathrm{Hom}\, C^o(X)$ carrying the initial topology induced by the families $\{ d'(f) \}_{f \in C(X)}$ and $\{ d'(f) \}_{f \in C^o(X)}$, respectively. Clearly the spaces just constructed are Hausdorff.

We leave it to the reader to verify that $\mathrm{Hom}_s C(X)$ is a subspace of $\mathrm{Hom}_s C^o(X)$.

How are $\mathrm{Hom}_s C(X)$ and $\mathrm{Hom}_s C^o(X)$ related to X ?

Every point $p \in X$ defines a homomorphism

$$i_X(p) \ : \ C(X) \longrightarrow \mathbb{R}$$

by requiring $i_X(p)(f) = f(p)$ for each $f \in C(X)$. Since we have $f(p) = d'(f)(i_X(p))$ for each $f \in C(X)$ and $p \in X$, the map

$$i_X \ : \ X \longrightarrow \mathrm{Hom}_s C(X),$$

sending each p into $i_X(p)$, is continuous.

The subset $i_X(X)$, turned into a subspace of $\mathrm{Hom}_s C(X)$, is homeomorphic to X_s. Indeed the map, assigning to each $i_X(p)$ the equivalence class $\pi(p) \in X_s$, is a bicontinuous bijection.

<u>Lemma 1</u>

For each convergence space X, the set $i_X(X)$ is dense in $\mathrm{Hom}_s C^o(X)$ and therefore is dense in $\mathrm{Hom}_s C(X)$.

<u>Proof:</u>

Let $h \in \mathrm{Hom}_s C^o(X)$. We have to show that for any choice of finitely many functions $f_1, \ldots, f_n$ in $C^o(X)$ and any positive real number ε the neighborhood U of h given by

$$\bigcap_{i=1}^{n} \{ k \in \mathrm{Hom}_s C^o(X) \mid \ | d'(f_i)(k) - d'(f_i)(h) | < \varepsilon \}$$

intersects $i_X(X)$ non-trivially.

Therefore let us form the function

$$g = \sum_{i=1}^{n} (f_i - h(f_i) \cdot \underline{1})^2 ,$$

for which $h(g) = o$. Clearly g is not a unit in $C^o(X)$ and hence assumes values which are arbitrarily close to zero. In other words, there is a $p \in X$ such that $g(p) < \varepsilon^2$. This means that

$$| f_i(p) - h(f_i) \cdot 1 | < \varepsilon$$

for $i = 1,\dots,n$, and hence $i_X(p) \in U$.

Lemma 1 yields immediately:

<u>Theorem 2</u>

For any convergence space X the map

$$i_X^* : C(\mathrm{Hom}_s C(X)) \longrightarrow C(X)$$

is an isomorphism whose inverse is

$$d' : C(X) \longrightarrow C(\mathrm{Hom}_s C(X)).$$

To represent $C^o(X)$ as the function algebra of $\mathrm{Hom}_s C^o(X)$ we first prove:

<u>Lemma 3</u>

For any convergence space X the space $\mathrm{Hom}_s C^o(X)$ is compact.

<u>Proof:</u>

Let a function $f \in C^o(X)$ be given. Since f is bounded, there is a natural number n such that $f(X) \subset [-n,n]$. Hence

$d'(f)(i_X(X)) \subset [-n,n]$, and because of Lemma 1 we have

$$d'(f)(\mathrm{Hom}_s C^o(X)) \subset [-n,n],$$

that is, $d'(f)$ is bounded for any function $f \in C^o(X)$.

Now if we give an ultrafilter Φ on $\mathrm{Hom}_s C^o(X)$, then clearly $d'(f)(\Phi)$ converges for any $f \in C^o(X)$. We define a map

$$h : C^o(X) \longrightarrow \mathbb{R},$$

assigning to each f the limit of $d'(f)(\Phi)$. This map is a homomorphism. Since $\mathrm{Hom}_s C^o(X)$ carries the initial topology induced by the family $\{ d'(f) \}_{f \in C^o(X)}$, we know that

$$\Phi \longrightarrow h \in \mathrm{Hom}_s C^o(X).$$

Hence $\mathrm{Hom}_s C^o(X)$ is compact.

The last two lemmas combined together immediately yield:

<u>Theorem 4</u>

For any convergence space X the space $\mathrm{Hom}_s C^o(X)$ is compact and

$$i_X^* : C(\mathrm{Hom}_s C^o(X)) \longrightarrow C^o(X)$$

is an isomorphism whose inverse is

$$d' : C^o(X) \longrightarrow C(\mathrm{Hom}_s C^o(X)).$$

Since for any convergence space X the space $\mathrm{Hom}_s C(X)$ is completely regular, we have that

$$i_X : X \longrightarrow \mathrm{Hom}_s C(X)$$

and therefore

$$i_X : X \longrightarrow \mathrm{Hom}_s C^o(X)$$

are homeomorphisms onto a subspace iff X is a completely regular topological space. Therefore we assume for the rest of this section that X is a completely regular topological space.

If we regard X as being identified with $i_X(X)$ via the map i_X, then we replace $\mathrm{Hom}_s C(X)$ and $\mathrm{Hom}_s C^o(X)$ by the symbols υX and βX respectively.

The spaces υX and βX are called the <u>realcompactification</u> and the <u>Stone-Čech-compactification</u> of X respectively.

Clearly υX is the largest subspace of βX to which every function in $C(X)$ has a continuous real-valued extension.

A space X is said to be <u>realcompact</u> iff $X = \upsilon X$. To give an example we point out that any subspace of $\mathbb{R}^n$ (n a natural number) is realcompact. For a very rich collection of examples of realcompact spaces we refer to [G,J].

For any compact space X we obviously have $X = \upsilon X = \beta X$. Clearly if X is not realcompact (resp. compact), then

$$d' : C(X) \longrightarrow C(\mathrm{Hom}_s C(X))$$

(resp. $d' : C^o(X) \longrightarrow C^o(\mathrm{Hom}_s C^o(X))$) are not induced by maps from $\mathrm{Hom}_s C(X)$ (respectively from $\mathrm{Hom}_s C^o(X)$) into X. This answers the question stated at the end of section (1.1).

Next let us derive the universal properties characterizing υX and βX. They are based on the fact that for any pair of convergence space Y and Z we have the following commutative diagram of continuous maps:

$$(1) \qquad \begin{array}{ccccc} Y & \xrightarrow{\;i_Y\;} & \mathrm{Hom}_s C(Y) & \subset & \mathrm{Hom}_s C^o(Y) \\ \Big\downarrow f & & \Big\downarrow f^{**} & & \Big\downarrow f^{**} \\ Z & \xrightarrow{\;i_Z\;} & \mathrm{Hom}_s C(Z) & \subset & \mathrm{Hom}_s C^o(Z) \end{array} ,$$

where f denotes a continuous map. By f^{**} we mean either the map sending each h into $h \circ f^* | C^o(Z)$ or $h \circ f^*$, depending upon whether $h \in \mathrm{Hom}_s C^o(Y)$ or $h \in \mathrm{Hom}_s C(Y)$.

Let X and T be completely regular topological spaces and $f : X \longrightarrow T$ be continuous. Via the identification mentioned above, diagram (1) turns into:

$$\begin{array}{ccccc} X & \subset & \upsilon X & \subset & \beta X \\ \downarrow{\scriptstyle f} & & \downarrow{\scriptstyle \upsilon f} & & \downarrow{\scriptstyle \beta f} \\ T & \subset & \upsilon T & \subset & \beta T \end{array} .$$

Here υf and βf are those maps extending f to υX and βX respectively.

We therefore have

<u>Theorem 5</u>

Any continuous map from a completely regular topological space X into a realcompact space can be (uniquely) extended to υX. This statement also holds true with realcompact replaced by compact, and υX replaced by βX. Hence if X is pseudocompact, i.e. if $C(X) = C^o(X)$, then $\upsilon X = \beta X$ and conversely.

We conclude the chapter by presenting an example of a completely regular topological space which is not realcompact. Let us denote by $W(\alpha)$ the set of all ordinals σ less than a given ordinal α. This set is well-ordered. We equip it with the interval topology. A system of basic open neighbourhoods of an element $\tau \in W(\alpha)$ consists of the sets of the form

$$(\sigma, \tau + 1) = \{ \lambda \mid \sigma < \lambda < \tau + 1 \} \qquad \sigma < \tau .$$

The space $W(\omega)$, where ω is the first infinite countable ordinal, is homeomorphic to $\mathbb{N}$, the set of natural numbers endowed with the discrete topology. For every ordinal α, the space $W(\alpha)$ is normal ([G,J], p.73). Now let ω_1 be the smallest uncountable ordinal. Then $W(\omega_1)$ is not compact. However, $W(\omega_1 + 1)$ is a compact space; indeed it is the one-point compactification of $W(\omega_1)$.

Call a space X countably compact if every family of closed sets with the finite intersection property has the countable intersection property or, equivalently, if every countable open cover has a finite refinement. A countably compact space is pseudocompact, as one easily verifies. Since every countable infinite subset of $W(\omega_1)$ has a limit point, $W(\omega_1)$ is countably compact, hence pseudocompact. Thus $W(\omega_1)$ is not realcompact. The Stone-Čech compactification of $W(\omega_1)$ is $W(\omega_1 + 1)$. For omitted details consult ([G,J], p. 73 ff.).

2. VECTOR SPACE TOPOLOGIES ON C(X) FOR WHICH THE EVALUATION MAP IS CONTINUOUS

As we learned in the previous chapter, two completely regular topological spaces need not be homeomorphic, even if their function algebras are isomorphic. But it is well known that any two completely regular topological spaces are homeomorphic iff their function algebras, endowed with either the topology of compact or of pointwise convergence are bicontinuously isomorphic [Mo,Wu].

However, the fact that both types of topologies on function algebras are in general not ω-admissible, together with the lack of completeness of these topologies (in general), forces us to devote some attention to vector space topologies on function algebras being at least ω-admissible.

We will use our knowledge about the most elementary facts of convergence spaces and of function algebras to derive that for a completely regular topological space X there is in general no ω-admissible vector space topology on $C(X)$.

The tool we will use is a so-called Marinescu structure [Ja] on $C(X)$. This structure - called I - turns out to be also useful in describing the relationship between X and $C_c(X)$. We therefore investigate I more closely than we would if it were necessary to use it only as a tool .

Throughout this chapter we denote by X a completely regular topological space.

2.1 A natural Marinescu structure on C(X)

The structure mentioned in the title relies on βX. This chapter let us reserve the symbol $\mathcal{K}$ for the collection of all compact subsets of $\beta X \setminus X$.

For any $K \in \mathcal{K}$ the space $\beta X \setminus K$ is a locally compact space containing X as dense subspace. The inclusion map $j_K : X \longrightarrow \beta X \setminus K$ induces a monomorphism

$$j_K^* : C(\beta X \setminus K) \longrightarrow C(X).$$

Let us therefore identify each $g \in C(\beta X \setminus K)$ with its restriction $g|X$, for any $K \in \mathcal{K}$. This means that for each $K \in \mathcal{K}$ the function algebra $C(\beta X \setminus K)$ is a lattice subalgebra of $C(X)$.

Basic to our construction is:

$$\bigsqcup_{K \in \mathcal{K}} C(\beta X \setminus K) = C(X).$$

To show this we must prove that each $g \in C(X)$ can be extended to $\beta X \setminus K$ for some $K \in \mathcal{K}$. Regarded g as a function into $\mathbb{R} \cup \{\infty\}$, the one-point-compactification of $\mathbb{R}$, the map g can be extended to βX. Clearly $g^{-1}(\infty) \subset \beta X \setminus X$ is a compact set (it might be empty). Hence

$g \in C(\beta X \setminus g^{-1}(\infty))$, establishing the above equality.

For each $K \in \mathcal{K}$ the continuous convergence structure on $C(\beta X \setminus K)$ coincides - because of the local compactness of $\beta X \setminus K$ - with the topology of compact convergence. Hence we have a family $\{ j_K^* \}_{K \in \mathcal{K}}$ of maps defined on topological algebras having their values in $C(X)$. The final convergence structure on $C(X)$ induced by the above family is denoted by I.

This convergence structure is a Marinescu structure in the sense of [Ja]. We will explain this term in the next section. The properties of I can be found in [Bi, Fe 1] and [Bi et al], out of which we take some of the following material.

The algebra $C(X)$ endowed with I, denoted by $C_I(X)$, is a convergence algebra [Bi, Fe 1], meaning that the algebra operations defined on $C_I(X) \times C_I(X)$ and $C_I(X) \times \mathbb{R}$ are continuous maps into $C_I(X)$. A filter Θ on $C(X)$ converges to zero in $C_I(X)$ iff Θ has as a basis a filter converging to zero in $C_c(\beta X \setminus K)$ for some $K \in \mathcal{K}$.

It is easy to show that I is ω-admissible. Hence

$$\mathrm{id} : C_I(X) \longrightarrow C_c(X)$$

is continuous. Let us point out that $C_c(X)$ is also a convergence algebra.

The two convergence algebras - $C_I(X)$ and $C_c(X)$ - are closely related to each other, as expressed in:

<u>Theorem 6</u>

Let X be a completely regular topological space. For any linear map t from a topological $\mathbb{R}$-vector space E into $C(X)$, the following are equivalent:

(i) t is continuous into $C_c(X)$

(ii) t is continuous into $C_c(\beta X \setminus K)$ for some $K \in \mathcal{K}$.

(iii) t is continuous into $C_I(X)$.

Proof:

We only show that (i) implies (ii) since the other implications are obvious.

The image filter $t(\mathfrak{U}(o))$ of the neighborhood filter $\mathfrak{U}(o)$ of zero in E converges to the zero function $\underline{o}$ in $C_c(X)$. Hence to each point $p \in X$ there is in X a neighborhood V_p of p and a filter element U_p in $\mathfrak{U}(o)$ for which

$$(2) \qquad \omega(t(U_p) \times V_p) \subset [\,-1,1\,].$$

The closure $cl_{\beta X} V_p$ (formed in βX) of V_p is in βX a neighborhood of p.

All the functions in $t(U_p)$, regarded as functions into $\mathbb{R} \cup \{\infty\}$, extend continuously to βX. They remain, however, real-valued on $cl_{\beta X} V_p$, because of (2). Since U_p is absorbant in E, all functions in $t(E)$ extend continuously and real-valued to $cl_{\beta X} V_p$.

Let us form in βX the interior W_p of $cl_{\beta X} V_p$. Clearly

$$\bigcup_{p\in X} W_p$$

is a locally compact subspace of βX and, as such, of the form $\beta X \setminus K$ for some $K \in \mathcal{K}$. From what we have just established we conclude:

$$t(E) \subset C(\bigcup_{p\in X} W_p) \subset C(X).$$

Next we prove the continuity of

$$t : E \longrightarrow C_c(\bigcup_{p\in X} W_p).$$

To any positive real number ε, we have

$$(3) \qquad \omega(t(\varepsilon \cdot U_p) \times W_p) \subset [\,-\varepsilon,\varepsilon\,]$$

for each $p \in X$. Since $\varepsilon \cdot U_p \in \mathfrak{U}(o)$ and since each $q \in \bigcup_{p\in X} W_p$

has a neighborhood of the form W_p, we deduce from (3) that $\omega(t(\mathcal{U}(o) \times \mathcal{W}(q))$ converges to $o \in \mathbb{R}$, where $\mathcal{W}(q)$ denotes the neighborhood filter in $\bigcup_{p \in X} W_p$ of q. Hence $t(\mathcal{U}(o))$ converges to $\underline{o} \in C_c(\bigcup_{p \in X} W_p)$.

2.2 A universal characterization of $C_I(X)$

The characterization of $C_I(X)$ mentioned in the title is based on the notion of the "inductive limit of topological vector spaces" taken in the category of convergence $\mathbb{R}$-vector spaces. By convergence $\mathbb{R}$-vector space we always mean an $\mathbb{R}$-vector space endowed with a convergence structure allowing the operations from $E \times E$ into E, and from $E \times \mathbb{R}$ into $\mathbb{R}$, to be continuous. The category of convergence $\mathbb{R}$-vector spaces is defined in the obvious way. Let $\{ E_\alpha \}_{\alpha \in N}$ be a family of topological $\mathbb{R}$-vector spaces such that:

(i) N is directed $(\alpha,\beta \in N \Rightarrow \exists\ \gamma \in N$ with $\gamma \geq \alpha,\beta)$.

(ii) to any two indices $\alpha,\beta \in N$ with $\beta \leq \alpha$ there is a continuous linear map

$$i_\beta^\alpha : E_\beta \longrightarrow E_\alpha .$$

$(E_\alpha)_{\alpha \in N}$ is called an <u>inductive family of topological $\mathbb{R}$-vector spaces</u>.

Let $\{ E_\alpha \}_{\alpha \in N}$ be an inductive family of topological $\mathbb{R}$-vector spaces. A convergence $\mathbb{R}$-vector space L is said to be the <u>inductive limit</u> of the family $\{ E_\alpha \}_{\alpha \in N}$ if the following two conditions hold:

(i') To any $\alpha \in N$ there is a continuous linear map

$$i_\alpha : E_\alpha \longrightarrow L,$$

such that $i_\alpha \circ i_\beta^\alpha = i_\beta$ for all $\alpha,\beta \in N$ with $\beta \leq \alpha$.

(ii') Let F be a convergence space. If to any $\alpha \in N$ there exists a continuous linear map $j_\alpha : E_\alpha \longrightarrow F$ such that for each $\beta \in N$ with $\beta \leq \alpha$ the diagram

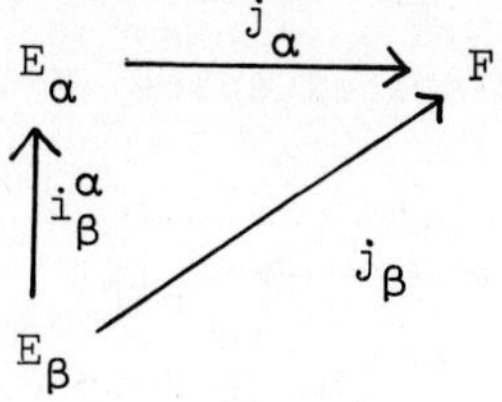

commutes, then there is a unique linear continuous map $j : L \longrightarrow F$ satisfying $j \circ i_\alpha = j_\alpha$ for each $\alpha \in N$.

We refer to [Fi] for the existence and uniqueness of the inductive limit (in the category of convergence $\mathbb{R}$-vector spaces) of a family of topological $\mathbb{R}$-vector spaces.

In case the family $\{ E_\alpha \}_{\alpha \in N}$ consists of locally convex vector spaces, its inductive limit is called a <u>Marinescu space</u> and the convergence structure on it a <u>Marinescu structure</u> [Ja] and [Ma].

The convergence $\mathbb{R}$-vector space $C_I(X)$ is the inductive limit of $\{ C_c(\beta X \setminus K) \}_{K \in \mathcal{K}}$: To any pair of compact sets $K, K' \subset \beta X \setminus X$ with $K \supset K'$ the maps

$$(j^K_{K'})^* : C_c(\beta X \setminus K') \longrightarrow C_c(\beta X \setminus K)$$

induced by the inclusion map

$$j^K_{K'} : \beta X \setminus K \longrightarrow \beta X \setminus K'$$

are obviously continuous. In fact because of the identification we made in the previous section, $(j^K_{K'})^*$ is the inclusion map. Moreover a linear map t from $C_I(X)$ into a convergence $\mathbb{R}$-vector space F is continuous iff

$$t \circ j_K^* : C_c(\beta X \setminus K) \longrightarrow F$$

is continuous for any $K \in \mathcal{K}$.

The property of $C_I(X)$, expressed in theorem 6, shall now be used to exhibit a certain class of convergence $\mathbb{R}$-vector spaces: A convergence $\mathbb{R}$-vector space F is said to satisfy property I_c with respect to a completely regular topological space X and a continuous linear map

$$i : F \longrightarrow C_c(X)$$

if any continuous linear map j from a topological $\mathbb{R}$-vector space E into $C_c(X)$ can be uniquely lifted to $\hat{j} : E \longrightarrow F$, which means that

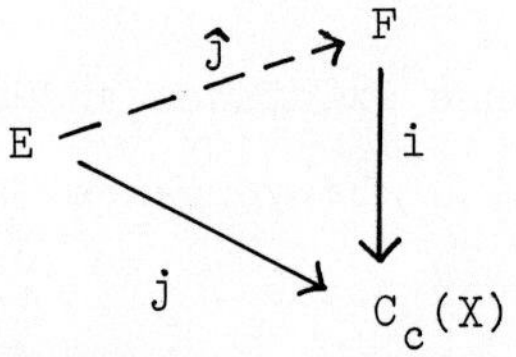

commutes.

Lemma 7

Let F be a convergence $\mathbb{R}$-vector space satisfying the property I_c with respect to a completely regular topological space X and the continuous linear map

$$i : F \longrightarrow C_c(X).$$

Then there exists one and only one continuous linear map

$$j : C_I(X) \longrightarrow F$$

for which $i \circ j = id$. Hence j is injective and i is surjective.

Proof:

For any compact set $K \subset \beta X \setminus X$ the inclusion map

$$j_K^* \;:\; C_c(\beta X \setminus K) \longrightarrow C_c(X)$$

is continuous. Hence it can be lifted in a unique way to

$$\hat{j}_K^* \;:\; C_c(\beta X \setminus K) \longrightarrow F.$$

For any pair of compact sets $K, K' \in \mathcal{K}$ with $K \supset K'$ we have a diagram of inclusions

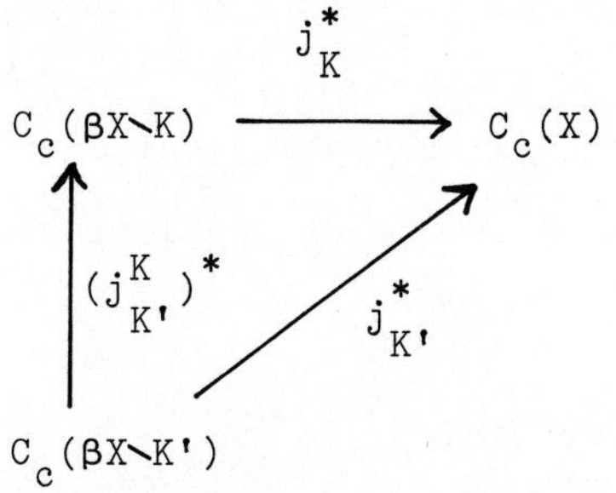

yielding the commutative diagram

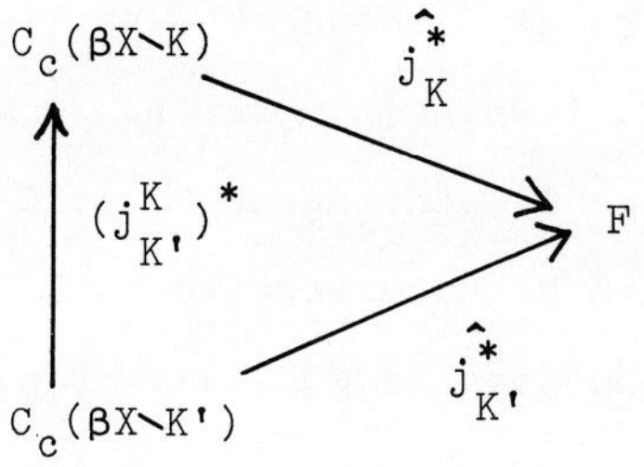

and thus a unique map j for which $i \circ j = \mathrm{id}$.

Using lemma 7 and theorem 6 the reader may verify as an exercise the following theorem [Bi et al]:

<u>Theorem 8</u>

Let L be a convergence $\mathbb{R}$-vector space satisfying the property I_c with respect to a completely regular topological space X and the continuous linear map

$$i : L \longrightarrow C_c(X).$$

If L is an inductive limit of a family of topological $\mathbb{R}$-vector spaces, then

$$i : L \longrightarrow C_I(X)$$

is a homeomorphism.

Corollary 9

Let X be a completely regular topological space. The algebra $C_c(X)$ is an inductive limit of a family of topological $\mathbb{R}$-vector spaces iff $C_c(X) = C_I(X)$

Let us point out that in general $C_c(X)$ and $C_I(X)$ do not coincide. To show this we form for each $n \in \mathbb{N}$ the set $F(I_n)$ of all functions in $C(\mathbb{Q})$ (where $\mathbb{Q}$ denotes the rationals) that vanish on I_n, the intersection of $[-n,n]$ with $\mathbb{Q}$. Obviously $\{F(I_n) | n \in \mathbb{N}\}$ forms a basis of a filter Θ convergent to $\underline{o} \in C_c(\mathbb{Q})$. In order that $\Theta \longrightarrow o \in C_I(\mathbb{Q})$ it would be necessary that $F(I_n) \subset C(\beta\mathbb{Q} \setminus K)$ for some compact set $K \subset \beta\mathbb{Q} \setminus \mathbb{Q}$ and some $n \in \mathbb{N}$. But for any point $p \in I_{n+1} \setminus I_n$ there is in $\beta\mathbb{Q}$ a neighborhood U not intersecting $cl_{\beta\mathbb{Q}} I_n \cup K$. Hence we would have a point $q \in U$ for which $f(q) = \infty$ and $f(I_n) = \{o\}$ for some $f \in C(X)$, contradicting $F(I_n) \subset C(\beta\mathbb{Q} \setminus K)$.

For necessary and sufficient conditions on X which yield $C_c(X) = C_I(X)$, see § 5.4.

2.3 ω-admissible vector space topologies on $C(X)$

Let T be an ω-admissible vector space topology on $C(X)$. The identity map

$$id : C_T(X) \longrightarrow C_c(X)$$

is continuous. Hence by theorem 6 we have $id(C(X)) = C(\beta X \setminus K)$ for some compact subset $K \subset \beta X \setminus X$. This is possible iff

$$X \subset \beta X \setminus K \subset \upsilon X,$$

which means that υX is in βX a neighborhood of X. Clearly if there is an open set W in βX satisfying

$$X \subset W \subset \upsilon X$$

then the topology on $C(X)$ of compact convergence on W is ω-admissible. Thus we have [Bi et al]:

Theorem 10

Given a completely regular space X there exists an ω-admissible $\mathbb{R}$-vector space topology on $C(X)$ iff υX is a neighborhood of X in βX. In case X is realcompact, an ω-admissible $\mathbb{R}$-vector space topology on $C(X)$ exists iff X is locally compact.

It is clear now that there exists no ω-admissible $\mathbb{R}$-vector space topology on $C(\mathbb{Q})$.

Let us call an ideal $J \subset C(Y)$, where Y denotes an arbitrarily given convergence space, <u>fixed</u> if there is a non-empty subset $A \subset X$ on which every function contained in J vanishes.

For any completely regular topological space Y a maximal ideal in $C_{co}(Y)$ is closed iff it is fixed [Mo,Wu]. Moreover, assigning to any closed maximal ideal the point at which it is fixed, we obtain a one-to-one correspondence between the set of all closed maximal ideals in $C_{co}(Y)$ and the points of X.

Given an ω-admissible $\mathbb{R}$-vector space topology T on $C(X)$ for which the closed maximal ideals are just the fixed ones, the continuity of

$$\mathrm{id} : C_T(X) \longrightarrow C_c(\beta X \setminus K)$$

for some compact set $K \subset \beta X \setminus X$ implies that $\beta X \setminus K = X$; hence X is locally compact. Since for a locally compact space X the topology of compact convergence on $C(X)$ is ω-admissible, we have:

<u>Theorem 11</u>

Let X be a completely regular topological space. There exists an ω-admissible $\mathbb{R}$-vector space topology on $C(X)$ for which the closed maximal ideals are just the fixed ones, iff X is locally compact.

In fact the local compactness of X can be characterized in various ways, some of them involving the existence of special ω-admissible topologies on $C(X)$:

<u>Theorem 12</u>

For any completely regular topological space X the following are equivalent:

(i) X is locally compact.

(ii) Among all the ω-admissible $\mathbb{R}$-vector space topologies on $C(X)$ there is one that is coarsest.

(iii) Among all the ω-admissible topologies on $C(X)$ there is one that is coarsest.

(iv) $C_c(X)$ carries the topology of compact convergence.

(v) $C_I(X)$ carries the topology of compact convergence.

(vi) $C_c(X)$ is topological

(vii) $C_I(X)$ is topological

Proof:

We will establish the following sequence of implications: (vi) $\Rightarrow$ (vii) $\Rightarrow$ (ii) $\Rightarrow$ (i) $\Rightarrow$ (iii) $\Rightarrow$ (i) $\Rightarrow$ (iv) $\Rightarrow$ (v) $\Rightarrow$ (vi).

(vi) implies by theorem 6 the identity of $C_c(X)$ and $C_I(X)$.

(vii) $\Rightarrow$ (ii): Since any ω-admissible $\mathbb{R}$-vector space topology on $C(X)$ is (by theorem 6) finer than I, (vii) implies that I is the coarsest of all ω-admissible $\mathbb{R}$-vector space topologies on $C(X)$.

(ii) $\Rightarrow$ (i): Let T be the coarsest of all ω-admissible $\mathbb{R}$-vector space topologies on $C(X)$. Clearly $id: C_T(X) \longrightarrow C_c(X)$ is continuous, implying, by theorem 6, that $C(X) = C(\beta X \setminus K)$ for some $K \in \mathcal{K}$ and that $id : C_T(X) \longrightarrow C_c(\beta X \setminus K)$ is continuous. Hence $C_T(X) = C_c(\beta X \setminus K)$. If X were strictly contained in $\beta X \setminus K$ there would be at least one point $p \in \beta X \setminus X$ not contained in K, such that $C_T(X) = C_c(\beta X \setminus (K \cup \{p\}))$. The homomorphism sending each $f \in C(X)$ into $f(p)$ is continuous on $C_c(\beta X \setminus K)$ but not on $C_c(\beta X \setminus (K \cup \{p\}))$. Hence $X = \beta X \setminus K$, meaning that X is locally compact.

(i) $\Rightarrow$ (iii): In case X is locally compact, the topology of compact convergence on $C(X)$ coincides with the continuous convergence structure.

(iii) $\Rightarrow$ (i): Let T be the coarsest of the ω-admissible topologies on $C(X)$. We will first show that the members of the neighborhood filter $\mathfrak{U}(o)$ of the zero function in $C_T(X)$ is absorbant. To this end let $f \in C(X)$. On $\mathbb{R} \cdot f \subset C(X)$ we choose the natural vector space topology. The final topology on $C(X)$ induced by the inclusion map is

ω-admissible. Hence the inclusion map of $\mathbb{R} \cdot f$ into $C_T(X)$ is continuous, implying that neighborhoods of zero absorb f. Since T is ω-admissible we find to each $p \in X$ a neighborhood V_p in X such that

$$\omega(U \times V_p) \subset [\,-1,1\,]$$

for some $U \in \mathfrak{U}(o)$. Hence all functions in U extend continuously to the interior W_p of $\mathrm{cl}_{\beta X} V_p$. Since $\mathfrak{U}(o)$ is absorbant all functions in $C(X)$ extend to $\bigcup_{p \in X} W_p$. Thus

$$C(X) = C(\bigcup_{p \in X} W_p).$$

For any real number $\lambda \neq o$, we consider the map m sending each $f \in C(X)$ into $\lambda \cdot f \in C(X)$. The final topology on $C(X)$ induced by the map

$$m : C_T(X) \times \{\lambda\} \longrightarrow C(X)$$

is ω-admissible. Hence $m : C_T(X) \times \{\lambda\} \longrightarrow C_T(X)$ is continuous. Moreover it is a homeomorphism. Thus for any $U \in \mathfrak{U}(o)$ the set $\lambda \circ U$ belongs to $\mathfrak{U}(o)$. Therefore for every positive real number ε, we have $\varepsilon \cdot U \in \mathfrak{U}(o)$, and thus we deduce that

$$\omega(\varepsilon \cdot U \times W_p) \subset [\,-\varepsilon,\varepsilon\,],$$

which implies the continuity of

$$id : C_T(X) \longrightarrow C_c(\bigcup_{p \in X} W_p)$$

at $\underline{0} \in C_T(X)$. Proceeding similarly as above, we deduce that for any $f \in C(X)$ different from $\underline{0}$, the map

$$C_T(X) \times \{f\} \longrightarrow C_T(X)$$

$$(g,f) \longrightarrow g + f$$

is a homeomorphism. This implies that

$$\mathrm{id} : C_T(X) \longrightarrow C_c(\bigcup_{p\in X} W_p)$$

is continuous. The assumption on T requires that

$$C_T(X) = C_c(\bigcup_{p\in X} W_p).$$

Reasoning as above we obtain the local compactness of X.

(i) ⇒ (iv): If X is locally compact, then $C_c(X)$ carries the topology of compact convergence.

(iv) ⇒ (v): If $C_c(X)$ carries the topology of compact convergence then, because of $C_c(X) = C_I(X)$, (theorem 6) $C_I(X)$ carries that topology too.

(v) ⇒ (vi): Theorem 6 yields the desired implication.

3. c-EMBEDDED SPACES

The relationship between a given completely regular topological space X and $C_T(X)$, where T denotes an $\mathbb{R}$-vector space topology, depends very much on T. At least from the technical point of view T should (aside from a functorial dependence on X) satisfy the following three conditions:

(i) T is ω-admissible

(ii) T is complete

(iii) T allows only the fixed maximal ideals to be closed.

As we pointed out in theorem 11 of the previous section, a topology T satisfying (i) and (iii) exists precisely when X is locally compact. If X is locally compact, then the topology of compact convergence on C(X) allows a very close relationship between X and $C_{co}(X)$.

Because of all this we use a convergence structure on C(X), namely the continuous convergence structure Λ_c (a natural generalization of the co-topology), and we show that it satisfies conditions analogous to (i) to (iii) above, stated for T. Thereby we allow X to be any convergence space.

Our next problem will consist of exhibiting a suitable class of convergence spaces, in which any two objects are homeomorphic iff their function algebras endowed with the continuous convergence structure are bicontinuously isomorphic. We call the spaces of this class c-embedded.

It will turn out that the property of c-embeddedness will be enjoyed by a large variety of topological spaces, especially by all completely regular ones.

Schroder's characterization of c-embedded spaces will make apparent how these spaces are built up by compact topological spaces.

We close the chapter by showing that the topology of compact convergence on $C(X)$, where X is c-embedded, is very closely related to Λ_c after all.

On this excursion, a technical convenience will be achieved by continuing investigations on $C_I(X_s)$ and deriving analoguous results as for $C_c(X)$. The reasons for choosing this method are not only the continuity of the maps

$$C_I(X_s) \xrightarrow{\text{id}} C_c(X_s) \xrightarrow{\Pi^*} C_c(X)$$

for some convergence space X, but also the fact that $C_I(X_s)$ is, for some of the problems we are interested in, much easier to handle than $C_c(X)$ itself.

Why we concentrate on the relationship between a c-embedded convergence space X and $C_c(X)$ rather than on the one between X and $C_I(X)$, (this object will be defined in the last section of this chapter too) is primarily because Λ_c is much coarser than I, and hence allows many more filters to converge, in other words, because more approximations are possible.

Throughout this chapter let us denote convergence spaces by X,Y,Z etc.

3.1 Completeness

We know that on $C(X)$ both convergence structures Λ_c and (in case X is a completely regular topological space) I are ω-admissible. In addition both are complete:

In a convergence $\mathbb{R}$-vector space G, a filter Θ is said to be a Cauchy filter if the filter $\Theta - \Theta$, generated by $\{T-T' \mid T, T' \in \Theta\}$, converges to zero. <u>Completeness</u> of G now means that any Cauchy filter converges to some element of G.

<u>Completeness of</u> $C_c(X)$: Let Θ be a Cauchy filter on $C_c(X)$ Since $\Theta - \Theta \longrightarrow \underline{o} \in C_c(X)$, the filter $\omega(\Theta \times \dot{p})$ is a Cauchy filter in $\mathbb{R}$ for any point $p \in X$. We denote its limit by $f(p)$. Assigning to each p this limit, we obtain a real-valued function f. The continuity of f can be shown as follows:

Since $\Theta - \Theta \longrightarrow \underline{o} \in C_c(X)$, to any point $p \in X$ and any filter Φ converging to $p \in X$, there are sets $T \in \Theta$ and $F \in \Phi$ with

$$\omega((T - T) \times F) \subset U,$$

for a given closed neighborhood U of $o \in \mathbb{R}$.

For any point $q \in F$ we have $\omega(\Theta \times \dot{q}) \longrightarrow f(q)$, which implies that $f(q)$ is adherent to $\{t(q) \mid t \in T\}$. Hence any number of $\{t(q) - f(q) \mid t \in T, q \in F\}$ is adherent to $\omega((T-T)\times F)$. Therefore

$$\{t(q) - f(q) \mid t \in T, q \in F\} \subset U.$$

Without loss of generality we may assume that $p \in F$, which yields:

$$\{(t(q) - f(q) - (t(p) - f(p)) \mid t \in T, q \in F\} \subset U - U$$

and therefore

$$f(p) - f(q) \in U - U + \{t(p) - t(q)\}$$

for any $q \in F$. Let $t \in T$ be fixed. The continuity of t allows us to choose $F' \in \Phi$ such that

$$(t(p) - t(q) \subset U \text{ for any } q \in F'.$$

Hence we have

$$(f(p) - f(q)) \in U - U + U$$

for any $q \in F \frown F'$. Choosing U such that $U\text{-}U\text{+}U \subset (-\varepsilon, \varepsilon)$ for a given positive real number ε, we end up with

$$|f(p) - f(q)| < \varepsilon \text{ for each } q \in F \frown F',$$

which states the continuity of f. The reader is left to show that $\Theta \longrightarrow f \in C_c(X)$.

Completeness of $C_I(X)$:

Let X now be a completely regular topological space. Assume Θ is a Cauchy filter on $C_I(X)$. There is a filter Ψ convergent to $\underline{o}$ in $C_c(\beta X \setminus K)$ (for some compact subset $K \subset \beta X \setminus X$) with the property that Ψ is a basis for $\Theta - \Theta$ in $C(X)$. Thus there is a set $M \in \Theta$ with $(M\text{-}M) \in \Psi$. We will show that M itself is in $C_c(\beta X \setminus K')$ for some compact $K' \subset \beta X \setminus X$. Let $g \in M$. For each $f \in M$ we have $(f\text{-}g) \in M\text{-}M$ and thus $(f\text{-}g) \in C(\beta X \setminus K)$. Hence

$$f^{-1}(\infty) \subset g^{-1}(\infty) \cup K.$$

Setting $g^{-1}(\infty) \cup K = K'$ we observe that Θ has a filter Θ' on $C(\beta X \setminus K')$ as a basis. Clearly Ψ is a basis for $\Theta' - \Theta'$. Since the inclusion map from $C_c(\beta X \setminus K)$ into $C_c(\beta X \setminus K')$ is continuous, $\Theta' - \Theta'$

is a Cauchy filter on $C_c(\beta X \setminus K')$ and hence converges there to some function t. Thus $\Theta \longrightarrow t \in C_I(X)$.

This proof can be found in [Bi,Fe 1]. For a general result on completeness of inductive limits see [Ja]. To summarize we state:

Theorem 13

Both convergence algebras $C_c(X)$ and (in case X is a completely regular topological space) $C_I(X)$ are complete.

As a contrast, let us take a look at $C_{co}(X)$ for a completely regular topological space X. The completion $\tilde{C}_{co}(X)$ of $C_{co}(X)$ consists of the set of all real-valued functions of X whose restrictions to all compact subspaces of X are continuous, and carries the topology of compact convergence. In case the compact sets are formed by finitely many points only, like in P-spaces ([G,J], p.63), then the completion of $C_{co}(X)$ consists of the collection of all real-valued functions of X, and carries the topology of pointwise convergence.

To present an example of a non-discrete P-space X we take a certain subspace of a space of the type $W(\alpha)$, defined at the end of § 1,2. Let α be the smallest ordinal of cardinality $\aleph_2$. Call it ω_2. Now let X be the subspace of $W(\omega_2)$ formed by deleting all non-isolated points having a countable basis. This space is not discrete and not realcompact ([G, J], p. 138)

However, for a large class of completely regular topological spaces, which contains all metric spaces, the associated algebras of all real-valued continuous functions endowed with the topology of compact convergence are complete [Wa].

As introduced in 1.1, we have for each subset A of an arbitrarily given convergence space the notion of the adherence $\bar{A}$ of A.

We call a set $A \subset Y$ dence in Y if $\bar{A} = Y$.

To conclude the section we prove:

Proposition 14

The $\mathbb{R}$-algebra $C^o(X)$, consisting of all bounded functions of $C(X)$, is dense in $C_c(X)$ and (in case X is a completely regular topological space) also in $C_I(X)$.

Proof:

Let Y be a completely regular topological space and $f \in C(Y)$. We form the sequence

$$\{(-\underline{n} \vee f) \wedge \underline{n}\}_{n \in \mathbb{N}},$$

of which all elements are bounded. Here $\underline{n}$ denotes the constant function assuming the value $n \in \mathbb{N}$. It is an easy exercise to verify that the Fréchet filter of the above sequence, which is the filter generated by all the sets of the form

$$F_i = \{ (-\underline{n} \vee f) \wedge \underline{n} \mid n > i \} \quad i \in \mathbb{N},$$

converges to f in $C_I(Y)$. Using the continuity of the maps

$$C_I(X_s) \xrightarrow{\text{id}} C_c(X_s) \xrightarrow{\Pi^*} C_c(X),$$

we deduce that $C^o(X)$ is dense in $C_c(X)$ for any convergence space.

3.2 Closed ideals

We now begin to exhibit a natural class of convergence spaces with the property that any two objects in it are homeomorphic iff their function algebras endowed with the continuous convergence structure are bicontinuously isomorphic.

Obviously any continuous map f from a convergence space X into a convergence space Y induces a continuous homomorphism

$$f^* : C_c(Y) \longrightarrow C_c(X).$$

Suppose now $k : C_c(Y) \longrightarrow C_c(X)$ is a continuous homomorphism. How can k induce some map from X into Y? To find out we form the set $\mathit{Hom}\, C_c(X)$, consisting of all real-valued continuous homomorphisms of $C_c(X)$ and endow this set with the continuous convergence structure, thereby obtaining $\mathit{Hom}_c C_c(X)$. We observe that for any $p \in X$

$$i_X(p) : C_c(X) \longrightarrow \mathbb{R}$$

is continuous and moreover that

$$i_X : X \longrightarrow \mathit{Hom}_c C_c(X)$$

is continuous too.

Returning to k we can check easily that

$$k^* : \mathit{Hom}_c C_c(X) \longrightarrow \mathit{Hom}_c C_c(Y),$$

sending each $h \in \mathit{Hom}_c C_c(X)$ into $h \circ k$, is a continuous map. Hence we have a continuous map

$$k^* \circ i_X : X \longrightarrow \mathit{Hom}_c C_c(Y).$$

In case Y were (via i_Y) homeomorphic to $\mathit{Hom}_c C_c(Y)$, we would have a continuous map f given by

$$i_Y^{-1} \circ k^* \circ i_X : X \longrightarrow Y,$$

for which $f^* = k$.

If in addition X were (via i_X) homeomorphic to $\mathit{Hom}_c C_c(X)$, then X and Y would be homeomorphic as soon as $C_c(X)$ and $C_c(Y)$ were bicontinuously isomorphic.

How restrictive is the condition that i_X is a homeomorphism? To find out, we are forced to study $Hom_c C_c(X)$. Clearly for any $h \in Hom_c C_c(X)$ the ideal $\text{Ker } h \subset C_c(X)$ is closed and maximal. We therefore investigate before our further development the closed ideals in $C_c(X)$.

Looking at the continuous maps

$$C_I(X_s) \xrightarrow{\text{id}} C_c(X_s) \xrightarrow{\Pi^*} C_c(X)$$

we observe that for any closed ideal $J \subset C_c(X)$ the ideal $(\Pi^*)^{-1}(J) \subset C_I(X_s)$ is closed too. We thus investigate first of all the closed ideals of $C_I(X)$, where X is assumed to be a completely regular topological space.

Evidently for every non-empty set $S \subset X$ the (proper) ideal $I(S)$ defined as

$$\{ f \in C(X) \mid f(S) = \{ o \} \}$$

is closed in $C_I(X)$. What we will establish [Bi,Fe 1] is:

Theorem 15

Let X be a completely regular topological space. For any ideal $J \subset C(X)$ let $N_X(J) \subset X$ be the collection of all points at which every function in J vanishes. An ideal $J \subset C_I(X)$ is closed iff $J = I(N_X(J))$. Hence a maximal ideal in $C_I(X)$ is closed if it consists of all functions in $C(X)$ that vanish on a single point.

Proof:

Let $J \subset C_I(X)$ be a closed ideal. We call $N_X(J) \subset X$ the null-set of J in X. Clearly

$$N_X(J) = \bigcap_{f \in J} Z(f),$$

where $Z(f)$ denotes the zero-set of f in X. Let J^o be the collection of all bounded elements in J. Since for any function $f \in J$ there is a bounded function $g \in J^o$ such that $Z(f) = Z(g)$, we can represent $N_X(J)$ as $\bigcap_{g \in J^o} Z_X(g)$. For this we refer to [G,J] 2.A, p.30. The inclusion map of X into βX induces a continuous map from $C_{co}(\beta X)$ into $C_I(X)$. Hence $J^o \subset C_{co}(\beta X)$ is closed, and is therefore of the form

$$J^o = I(N_{\beta X}(J^o)),$$

where $N_{\beta X}(J^o)$ is a closed non-empty subset of βX. First of all we will verify that J^o contains all bounded functions of $I(N_X(J))$. We are obviously done as soon as we know that $N_{\beta X}(J^o)$ is the closure in βX of $N_X(J)$. Assume to the contrary that $N_{\beta X}(J^o)$ contains a point q outside of $cl_{\beta X} N_X(J)$. We choose in βX a closed neighborhood U for q disjoint from $cl_{\beta X} N_X(J)$. There is a function $g \in C(\beta X)$ such that $g(q) = 1$ and g vanishes outside of U. We assert that $g \in J \cap C(\beta X \setminus K)$ where K denotes the compact set $U \cap N_{\beta X}(J^o) \subset \beta X \setminus X$. Clearly $J \cap C_{co}(\beta X \setminus K)$ is a closed ideal in $C_{co}(\beta X \setminus K)$, and therefore consists of all functions in $C(\beta X \setminus K)$ that vanish on the null set in $\beta X \setminus K$ of $J \cap C_{co}(\beta X \setminus K)$. Since the bounded functions in $J \cap C(\beta X \setminus K)$ are precisely the elements of J^o, we conclude that $N_{\beta X}(J^o) \cap (\beta X \setminus K)$ is the null set of $J \cap C(\beta X \setminus K)$. The function g vanishes on $N_{\beta X}(J^o) \cap (\beta X \setminus K)$ and therefore g is an element of $J \cap C(\beta X \setminus K)$, as claimed. Thus we know $g \in J^o$. On the other hand g is not an element of $I(N_{\beta X}(J^o))$, which is of course J^o. Because of this contradiction, we conclude that $N_{\beta X}(J^o) = cl_{\beta X} N_X(J)$. Hence $N_X(J) \neq \emptyset$, and J^o consists of all bounded functions in $I(N_X(J))$. Let $f \in I(N_X(J))$. There is a unit $u \in C(X)$ such that $u \cdot f$ is bounded. Hence $u \cdot f \in J^o$ and thus $f \in J$. This implies that $J = I(N_X(J))$. The proof of the converse is evident.

Because of the complete regularity of X we moreover have:

Corollary 16

Let X be a completely regular topological space. The closed ideals in $C_I(X)$ are in a one-to-one correspondence with the closed non-empty subsets of X.

Let X now be a convergence space. Let us call an ideal in $C(X)$ full if it consists of all functions vanishing on a fixed closed non-empty subset of X. Obviously any full ideal in $C_c(X)$ is closed. Moreover for a full ideal $J \subset C(X)$ we have $J = I(N_X(J))$.

From theorem 15 and the continuity of the maps

$$C_I(X_s) \xrightarrow{\text{id}} C_c(X_s) \xrightarrow{\Pi^*} C_c(X),$$

one easily deduces the following two theorems [Bi 3]:

Theorem 17

For any convergence space X an ideal in $C_c(X)$ is closed iff it is full. Hence a maximal ideal in $C_c(X)$ is closed iff it is fixed.

This theorem leads us to a very fundamental result of our study:

Corollary 18

For any convergence space X the map $i_X : X \longrightarrow \mathit{Hom}_c C_c(X)$ is a continuous surjection.

Theorem 19

For any convergence space X the closed ideals in $C_c(X)$ are, via the forming of null sets, in a one-to-one correspondence with the closed non-empty subsets of the completely regular topological space X_s associated to X.

In addition the reader may verify that X_s and $\mathit{Hom}_c\, C_I(X_s)$ are homeomorphic via i_{X_s}.

Let us refer to [Bi 3] and [Bi 2] for the study of the adherence of an ideal in $C_c(X)$ and another proof of corollary 18.

3.3 c-embeddedness

In this section we will turn our attention to those spaces X for which i_X is a homeomorphism. We call these spaces c-embedded spaces [Bi 1] , [Bi 2]. A characterization of the c-embedded topological spaces ensures us that the condition of c-embeddedness is not too restrictive. Moreover we will see that the class of all c-embedded spaces is big enough to reproduce all $C_c(X)$, where X runs through all convergence spaces. We close the section by a result which shows that each c-embedded convergence space X is determined by $C_c(X)$.

We begin by collecting some basic properties of c-embeddedness. Since $C_c(Y)$ is Hausdorff for any Y, any c-embedded space is Hausdorff.

Call a convergence space X regular if it is Hausdorff and for each point $p \in X$ and each filter Φ convergent to p, the filter $\bar{\Phi}$, generated by $\{\bar{F} \mid F \in \Phi\}$ converges to p. It is easy to see that $C_c(Y)$ is regular for any convergence space Y. Since $\mathit{Hom}_c C_c(X)$

is a closed subspace of $C_c(C_c(X))$, it is regular. Hence every c-embedded space has to be regular. Note that regularity of a space X is not a consequence of the injectivity of i_X. For the space S introduced in § 1.1 is not regular; however i_S is injective.

We leave it to the reader to verify the following permanence properties:

<u>Proposition 20</u>

Any subspace of a c-embedded space is c-embedded. The cartesian product of any family of c-embedded spaces is c-embedded.

Basic to our further techniques is the following:

<u>Lemma 21</u>

For any convergence space X the convergence spaces $C_c(X)$ and $Hom_c C_c(X)$ are c-embedded.

These results follow immediately from the diagram

$$\begin{array}{ccc} C_c(X) & \xrightarrow{\ i_{C(X)}\ } & C_c(C_c(C_c(X))) \\ & \searrow^{\mathrm{id}} & \downarrow {\scriptstyle i_X^{*}} \\ & & C_c(X) \ , \end{array}$$

where i_X is regarded as a map from X into $C_c(C_c(X))$, and proposition 20.

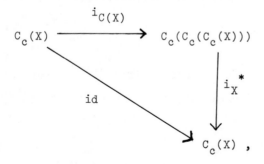

To see that not every c-embedded space is topological, choose a space Y such that $C_c(Y)$ is not topological (theorem 12). Let us start the more detailed studies of c-embedded spaces by investigating the topological ones:

For any completely regular topological space X the map $i_X : X \longrightarrow \mathit{Hom}_s C_c(X) \subset \mathrm{Hom}_s C(X)$ is a homeomorphism. Since $id : \mathit{Hom}_c C_c(X) \longrightarrow \mathit{Hom}_s C_c(X)$ is continuous we have:

Lemma 22

Any completely regular topological space is c-embedded.

However, completely regular topological spaces are not the only topological spaces that are c-embedded. For the formulation of the characteristic property which c-embedded topological spaces enjoy, we need the following notion:

Let us call a subset S of a convergence space X weakly closed if $S = \Pi^{-1}(\overline{\Pi(S)})$, where $\Pi : X \longrightarrow X_s$ denotes the canonical projection.

Now let us state B. Müller's characterization [Mü] :

Theorem 23

A topological space X is c-embedded iff $C(X)$ separates points and every neighborhood filter has a basis consisting of weakly closed sets.

Proof:

It is easy to verify that c-embedded topological spaces satisfy the condition expressed in the proposition. To show the converse, we copy the corresponding proof from [Bu, Mü].

Since $C(X)$ separates points in X, the map

$$i_X : X \longrightarrow \mathit{Hom}_c C_c(X)$$

is a continuous bijection. To show that it is a homeomorphism, we choose an arbitrarily given filter Φ convergent to $h \in \mathit{Hom}_c C_c(X)$, and we will demonstrate that $i_X^{-1}(\Phi) \longrightarrow i_X^{-1}(h) \in X$.

Instead of $i_X^{-1}(h)$ we write p. To any point $q \in X$ we choose a weakly closed neighborhood U_q such that $p \notin U_q$ whenever $p \neq q$.

The filter Φ generated by the collection

$$\{ I(U_q) \mid q \in X \}$$

of ideals converges to $\underline{0}$ in $C_c(X)$. Hence we find $F \in \Theta$ and $V \in i_X^{-1}(\Phi)$ such that

$$F(V) \subset [-1,1].$$

Clearly F contains $\bigcap_{i=1}^{n} I(U_{q_i})$ for some points $q_1,\dots,q_n \in X$.

Let us verify now that

$$V \subset U_p \cup U_{q_1} \cup \dots \cup U_{q_n}.$$

Assume $q \notin (U_p \cup \bigcup_{i=1}^{n} U_{q_i})$. We find $f \in C(X)$, assuming the value 2 at q and vanishing on $U_p \cup \bigcup_{i=1}^{n} U_{q_i}$. Hence

$$f \in \bigcap_{i=1}^{n} I(U_{q_i}) \subset F,$$

and therefore $f(V) \subset [-1,1]$. Thus $q \notin V$, establishing the above inclusion. Since Π is injective, it is the identity on X. Obviously $i_X^{-1}(\Phi) \longrightarrow p \in X_s$. Because of $p \notin \bigcup_{i=1}^{n} U_{q_i}$ the filter $i_X^{-1}(\Phi)$ contains the complement W of $\bigcup_{i=1}^{n} U_{q_i}$ in X_s. Since $U_p \cup \bigcup_{i=1}^{n} U_{q_i}$ and W are members of $i_X^{-1}(\Phi)$ we conclude from

$$U_p \supset W \cap (U_p \cup \bigcup_{i=1}^{n} U_{q_i})$$

that $U_p \in i_X^{-1}(\Phi)$.

The weakly closed neighborhood U_p was arbitrarily choosen. Hence $i_X^{-1}(\Phi)$ is finer than the neighborhood filter of $p \in X$. Thus $i_X^{-1}(\Phi)$ converges to $p \in X$.

In fact every c-embedded topological space X is regular but not necessarily completely regular. On the other hand, a regular topological space need not be c-embedded.

Let us present two examples, stated in [Bu,Mü], which show that the notion of c-embeddedness for topological spaces is intermediate between those of complete regularity and regularity in combination with the separation of points by real-valued continuous functions. Both examples are based on the Tychonoff plank T. This space is defined as follows. Consider $W(\omega)$ and $W(\omega_1)$ as introduced in § 1.2 and its one-point compactifications $W(\omega + 1)$ and $W(\omega_1 + 1)$ respectively. Then T is the space

$$W(\omega + 1) \times W(\omega_1 + 1) \setminus \{(\omega,\omega_1)\} .$$

(The Stone-Čech compactification of T is $W(\omega + 1) \times W(\omega_1 + 1)$. For further interesting details on T consult [G,J]). For each $n \in \mathbb{N}$ set $T_n = T \times \{ n \}$ and form the topological sum $\sum_{n \in Z} T_n$. In the ladder space we identify

$$(p, \omega_1, 2k) \text{ with } (p, \omega_1, 2k + 1)$$

and

$$(\omega, q, 2k + 1) \text{ with } (\omega, q, 2k + 2),$$

for all $k \in \mathbb{Z}$, $p \in W(\omega)$ and $q \in W(\omega_1)$. We equip the resulting set of equivalence classes with the quotient topology and denote it by S. Let $P = S \cup \{ t \}$ where $t \notin S$. Define a topology on P as follows: A neighborhood base in P of any point $q \in S$ is the filter of neighborhoods in S of q. A neighborhood base of t in P is given by

$$\{ \{t\} \cup \bigcup_{\nu \leq n} \tilde{T}_\nu \mid n \in \mathbb{Z} \}$$

where $\tilde{T}_\nu$ is the image in S of T_ν. The space P is regular and c-embedded, but not completely regular.

For the next example set $P_n = P \times \{ n \}$ for each $n \in \mathbb{N}$ and form the topological sum $\sum_{n \in \mathbb{Z}} P_n$. To this sum we again adjoin one point, say $\hat{t} \notin \sum_{n \in \mathbb{Z}} P_n$. We introduce a topology on

$$R = \{ \hat{t} \} \cup \sum_{n \in \mathbb{Z}} P_n$$

as follows: A neighborhood base in R of any point $q \in \sum_{n \in \mathbb{Z}} P_n$ is the filter of neighborhoods in $\sum_{n \in \mathbb{Z}} P_n$ of q. A neighborhood base of $\hat{t}$ in R is given by

$$\{ \{\hat{t}\} \cup \bigcup_{\substack{\nu \geq n \\ \mu \geq n}} T_\nu \times \{ \mu \} \mid n \in \mathbb{N} \}.$$

The space R is regular and admits point-separating continuous real-valued functions; however, it is not c- embedded.

As we demonstrated so far the class of c-embedded convergence spaces contains a large collection of topological spaces. It will become apparent later that the collection of non-topological c-embedded convergence spaces is also very big.

To conclude this paragraph we will show that any $C_c(X)$ is bicontinuously isomorphic to $C_c(Hom_c C_c(X))$ and we will state a precise statement of the fact that any c-embedded X is determined by $C_c(X)$.

Theorem 24

For any convergence space X the map

$$i_X^* : C_c(\mathit{Hom}_c C_c(X)) \longrightarrow C_c(X)$$

is a bicontinuous isomorphism whose inverse is

$$d : C_c(X) \longrightarrow C_c(\mathit{Hom}_c C_c(X)),$$

defined by $d(f)(h) = h(f)$ for any $f \in C(X)$ and any $h \in \mathit{Hom}_c C_c(X)$.

Proof:

Since i_X^* is injective and $i_X^* \circ d = id_{C(X)}$ holds, d is an isomorphism.

Theorem 25

Any continuous map $f : X \longrightarrow Y$ induces a continuous homomorphism $f^* : C_c(Y) \longrightarrow C_c(X)$. If Y is c-embedded any continuous homomorphism $k : C_c(Y) \longrightarrow C_c(X)$ induces a continuous map $k^* : \mathit{Hom}_c C_c(X) \longrightarrow \mathit{Hom}_c C_c(Y)$ for which $(i_Y^{-1} \circ k^* \circ i_X)^* = k$. In case both X and Y are c-embedded, k is a bicontinuous isomorphism iff $i_Y^{-1} \circ k^* \circ i_X$ is a homeomorphism. Hence, two c-embedded spaces X and Y are homeomorphic iff $C_c(X)$ and $C_c(Y)$ are bicontinuously isomorphic.

The proof is obvious.

3.4 Compact and locally compact c-embedded spaces

To prepare a general characterization of c-embedded spaces in terms of certain limits of compact topological spaces, we will study first compact and locally compact convergence spaces.

We call a convergence space X compact if it is Hausdorff and every ultrafilter on X converges in X.

Note that not every compact convergence space is topological. Instead of presenting a particular example we refer to [Ra, Wy] and [Wo] for the study of completions and compactifications of so-called Cauchy spaces.

A set F in a convergence space X is called compact if F is compact when regarded as a subspace. A convergence space is said to be locally compact if it is Hausdorff and every convergent filter contains a compact set. As in topology compactness for convergence spaces can be characterized in terms of "coverings".

A system $\mathcal{S}$ of subsets of a convergence space X is called a covering system if each convergent filter on X contains some element of $\mathcal{S}$.

To complete the notions which we are going to use in our description of compact spaces (see [Sch 1]), let us introduce: A point $p \in X$ is called adherent to a filter Φ on X or a point of adherence of Φ if p is the limit of a filter Ψ (i.e. $\Psi \longrightarrow p$) finer than Φ. Clearly any limit of a filter Φ on X is adherent to Φ.

Proposition 26

Let X be a Hausdorff convergence space. The following are equivalent:

(i) X is compact.

(ii) Every filter on X has a point of adherence.

(iii) In every covering system there are finitely many members, of which the union is X.

Proof:

(i) $\Rightarrow$ (ii): By Zorn's lemma, to every filter Φ on X there is an ultrafilter Ψ on X with $\Psi \geqslant \Phi$. Any limit of Ψ is adherent to Φ.

(ii) ⇒ (iii): Let $\mathcal{S}$ be a covering system allowing no finite subcover. Hence $\{ X \setminus S \mid S \in \mathcal{S} \}$ generates a filter, say Φ, on X. Every convergent filter in X has a basis in some member of $\mathcal{S}$. Thus Φ does not have an adherent point in X.

(iii) ⇒ (i): Assume that some ultrafilter on X, say Φ does not converge in X. Then Φ cannot be finer than any convergent filter Ψ. Since for any subset $M \subset X$ either M or $X \setminus M$ belongs to Φ, we find in any convergent filter Ψ a member $M_\Psi \in \Psi$ for which $X \setminus M_\Psi$ belongs to Φ. The system $\{ M_\Psi \mid \Psi \text{ convergent in } X \}$ is a covering system of X. If finitely many members of this system would cover X, then Ψ would have to contain the empty set.

Let X now be a compact convergence space. Using ultrafilters one proves, as in general topology, that for any continuous map f from X into a Hausdorff convergence space Y the image $f(X)$ is compact; thus any continuous real-valued function of X is bounded.

Moreover Λ_c on $C(X)$ is nothing but the supremum norm topology. We obtain a guide for our intuition for the forth-coming problems by taking a quick look at a compact space X. The isomorphism

$$\Pi^* : C_c(X_s) \longrightarrow C_c(X),$$

induced by $\Pi : X \longrightarrow X_s$, is bicontinuous, since both algebras carry the uniform topology. Hence, if X is c-embedded it is homeomorphic to X_s, by theorem 25, and thus a Hausdorff topological space.

Proposition 20 yields therefore:

<u>Proposition 27</u>

Any compact subspace of a c-embedded space is topological.

In case X is a locally compact space, then $C_c(X)$ carries the initial topology generated by all sup-seminorms, that is by seminorms of type

$$\begin{array}{ccc} C(X) & \xrightarrow{s_K} & \mathbb{R} \\ \cup\!\!\!\!\!\!\!\!\;\;\,\in & & \cup\!\!\!\!\!\!\!\!\;\;\,\in \\ f & \longrightarrow & \sup_{p\in K}|f(p)| \end{array} ,$$

where $K \subset X$ is a compact subset of X. Thus $C_c(X)$ carries in this case the topology of compact convergence.

To prepare for the study of locally compact spaces we first focus on $\mathcal{L}_c E$, the dual space $\mathcal{L}E$ of a topological $\mathbb{R}$-vector space E endowed with the continuous convergence structure (see [Sch 1]).

Proposition 28

Let E be a topological $\mathbb{R}$-vector space. For any neighborhood U of zero the polar U^o is a compact topological subspace of $\mathcal{L}_c E$; in fact it carries the topology of pointwise convergence.

Proof:

Let Φ be an ultrafilter on U^o. Since U^o_s, the polar U^o endowed with the topology of pointwise convergence, is compact Φ converges in U^o_s to some functional f. We proceed to show that Φ even converges to f in $\mathcal{L}_c E$.

For any element $e \in E$ and any positive real number ε we find a $T \in \Phi$ with

$$\omega(T \times \{e\}) \subset f(e) + [-\frac{\varepsilon}{2}, \frac{\varepsilon}{2}].$$

Hence for any $k \in T$ we have

$$k(e + \frac{\varepsilon}{2} \cdot U) \subset k(e) + \frac{\varepsilon}{2} \cdot k(U) \subset f(e) + [-\frac{\varepsilon}{2}, \frac{\varepsilon}{2}] + \frac{\varepsilon}{2} \cdot [-1,1].$$

But this shows that

$$\omega(T \times (e + \frac{\varepsilon}{2} \circ U)) \subset f(e) + [-\varepsilon, \varepsilon],$$

as desired. Since $\mathcal{L}_c E$ is a c-embedded space U_c^o, the set U^o endowed with the continuous convergence structure, is a compact c-embedded space (Proposition 20). By the above remarks it is thus topological and hence homeomorphic to U_s^o.

Corollary 29

For any topological $\mathbb{R}$-vector space E the c-dual $\mathcal{L}_c E$ is locally compact. Each compact set in $\mathcal{L}_c E$ is topological; it carries the topology of pointwise convergence. Hence, for any topological unitary $\mathbb{R}$-algebra A the space $Hom_c A$, consisting of $Hom\, A$, the set of all real-valued unitary homomorphisms of A, together with the continuous convergence structure, is locally compact.

Proof:

Since any filter convergent to zero in $\mathcal{L}_c E$ contains the polar of a neighborhood of zero in E, the first part of the corollary follows immediately. Any compact subspace of $\mathcal{L}_c E$ is c-embedded and thus is topological, hence it carries the topology of pointwise convergence. Evidently $Hom\, A \subset \mathcal{L}_c A$ is a closed subset and therefore $Hom_c A$ is locally compact.

Let us point out that $\mathcal{L}_c E$ is not topological unless it is finite dimensional.

Next suppose A is a normed $\mathbb{R}$-algebra. We may assume that the norm $\| \ \|$ of A satisfies $\|a \cdot b\| \leq \|a\| \cdot \|b\|$ for any choice of $a, b \in A$, and let us suppose in addition that the identity has norm one.

Note that any homomorphism in *Hom* A has norm less than or equal to 1, see [Ri], meaning that *Hom* A is in the polar of the unit ball of A. This means:

Corollary 30

If A is a normed $\mathbb{R}$-algebra then $Hom_c A$ is a compact topological space, which carries the topology of pointwise convergence.

To collect the results on compact c-embedded spaces we state [Bi 4]:

Theorem 31

For any c-embedded convergence space X the following are equivalent:

(i) X is compact.

(ii) X is compact and topological.

(iii) $C_c(X)$ carries a norm topology.

If $C_c(X)$ carries a norm topology, then it is the topology of uniform convergence.

Proof:

The implication (iii) ⇒ (ii) is a consequence of corollary 30, and the rest is well known.

If for a c-embedded space X the convergence algebra $C_c(X)$ is topological, then X is by corollary 29 a locally compact space. Thus locally compact c-embedded spaces can be described by their associated convergence function algebra as expressed by M. Schroder in [Sch.1]:

Theorem 32

For any c-embedded space X the following are equivalent:

(i) X is locally compact.

(ii) $C_c(X)$ is topological.

If $C_c(X)$ carries a topology, then it is the topology of compact convergence.

Let us now investigate locally compact spaces from another point of view. We would like to characterize locally compact c-embedded spaces by topological properties.

For this purpose we first generalize our notion of an inductive limit, defined in section 2.2., to general convergence spaces. All we really need to do is to require that in the introduction of the inductive limit of topological vector spaces in section 2.2., the spaces and the maps appearing there be replaced by convergence spaces and continuous maps respectively.

Any locally compact space X is the inductive limit of the family $\mathcal{K}$ of all compact subspaces of X directed by inclusion. Clearly if X is in addition c-embedded every member of $\mathcal{K}$ is topological (proposition 27). Moreover $C(X)$ separates points of X.

Conversely assume that X is the inductive limit of an inclusion-directed family $\{ K_\alpha \}_{\alpha \in N}$ of compact topological subspaces. Assume in addition that $C(X)$ separates points of X. Clearly X is a locally compact space which we will show is c-embedded. Evidently

$$i_X : X \longrightarrow \mathcal{H}om_c C_c(X)$$

is a continuous bijection.

Hence $i_X | K_\alpha$ is a homeomorphism from K_α onto $i_X(K_\alpha)$ for each

$\alpha \in N$. We will show now that each compact set $H \subset Hom_c\, C_c(X)$ is contained in $i_X(K_\alpha)$ for some $\alpha \in N$. This will imply that i_X is a homeomorphism since $Hom_c C_c(X)$ carries the final convergence structure induced by the family of all compact subspaces.

Since the $\{ K_\alpha \}_{\alpha \in N}$ forms a covering system of X, the topology of $C_c(X)$ is determined by the collection of seminorms

$$s_\alpha : C(X) \longrightarrow \mathbb{R},$$

assuming on each $f \in C(X)$ the value $\sup_{p \in K_\alpha} |f(p)|$, where α varies over N. Since $Hom_c C_c(X)$ is locally compact, the topology of $C_c(Hom_c C_c(X))$ is determined by all seminorms of the form

$$s_H : C(Hom_c C_c(X)) \longrightarrow \mathbb{R},$$

sending each $g \in C(Hom_c C_c(X))$ into $\sup |g(h)|$, taken over a compact set $H \subset Hom_c C_c(X)$.

Since $i_X^* : C_c(Hom_c C_c(X)) \longrightarrow C_c(X)$ is a bicontinuous isomorphism, $s_H \circ (i_X^*)^{-1}$ (called s_H') is a continuous seminorm on $C_c(X)$. Hence

$$s_H' \leq r \cdot s_\alpha$$

for some positive real number r and some $\alpha \in N$. The kernels (the zero sets) of s_H' and s_α are related by

$$\ker(s_H') \supset \ker(s_\alpha).$$

The null sets of these two closed ideals are therefore related by

$$N_X(\ker(s_H')) \subset N_X(\ker(s_\alpha)).$$

Since $(i_X^*)^{-1}(\ker(s_H'))$ and $(i_X^*)^{-1}(\ker(s_\alpha))$ are closed ideals in $C_c(Hom_c C_c(X))$, with H and $i_X(K_\alpha)$ as their null sets respectively, we have $H \subset i_X(K_\alpha)$. Therefore i_X is a homeomorphism. Thus we have:

Theorem 33

A convergence space X is locally compact and c-embedded iff the following two conditions hold:

(i) $C(X)$ separates the points of X.

(ii) X is the inductive limit of a family of compact topological subspaces directed by inclusion.

Hence a locally compact convergence space X is c-embedded iff $C(X)$ separates points and in addition every compact subspace is topological.

Our example in § 0.1 of a non-topological convergence space, the space E, is evidently locally compact. It is also c-embedded. Indeed, since every linear map from E into $\mathbb{R}$ is continuous, $C(E)$ separates the points of E. Using proposition 26, one verifies that every compact subset of E is contained in a finite dimensional subspace of E. The finite dimensional subspaces of E carry the natural topology. Thus every compact subset of E is topological. Hence, by the above theorem, E is c-embedded.

3.5 Characterization of c-embedded spaces

Any completely regular topological space Y is the projective limit of all locally compact subspaces of βY that contain Y. Analogously we will represent any c-embedded space X as a projective limit of locally compact c-embedded spaces. We use here an idea of a preprint [Sch 2] of M. Schroder, to whom this characterization is due.

To introduce the notion of the <u>projective limit</u> let $\{Y_\alpha\}_{\alpha\in M}$ be a family of convergence spaces indexed by a downward directed set M. We call $\{Y_\alpha\}_{\alpha\in M}$ a <u>projective family</u>, if for any $\alpha,\beta \in M$ with $\alpha \leqslant \beta$ there is a continuous map

$$i^\alpha_\beta : Y_\alpha \longrightarrow Y_\beta .$$

The projective limit $\operatorname{proj}_{\alpha\in M} Y_\alpha$ of a projective family $\{Y_\alpha\}_{\alpha\in M}$, or in shorter form the projective limit of $\{Y_\alpha\}_{\alpha\in M}$, is a convergence space L with the following two properties:

(i) To each $\alpha \in M$ there is a continuous map

$$\pi_\alpha : L \longrightarrow Y_\alpha$$

with $i^\alpha_\beta \circ \Pi_\alpha = \Pi_\beta$ for each pair $\alpha,\beta \in M$ with $\alpha \leqslant \beta$.

(ii) In case there is to each $\alpha \in M$ a continuous map j_α from a fixed convergence space F into Y_α such that

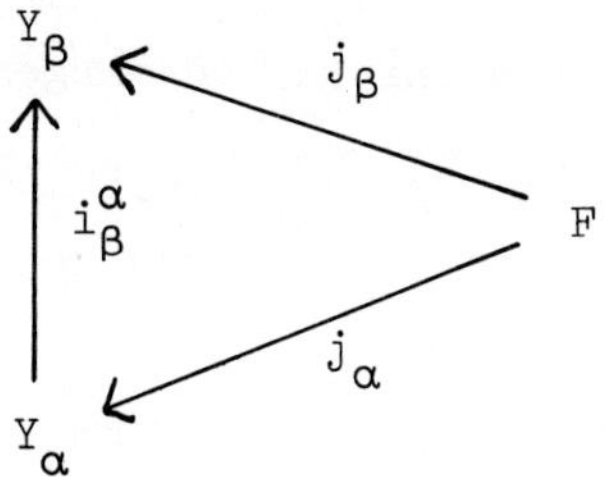

commutes for any choice of $\alpha,\beta \in M$ with $\alpha \leqslant \beta$, then there is a unique continuous map $j : F \longrightarrow L$ with

$$\pi_\alpha \circ j = j_\alpha .$$

Obviously if L exists, then it is determined up to homeomorphism.

To show the existence of the projective limit, let us form $\prod_{\alpha\in M} Y_\alpha$. The subspace of all those elements $g \in \prod_{\alpha\in M} Y_\alpha$ for which

$i^{\alpha}_{\beta}(\Pi_{\alpha}(g)) = \Pi_{\beta}(g)$ for any choice of α,β with $\alpha \leq \beta$, is a model of L, where Π_{α} and Π_{β} denote the canonical projections.

If each member of the family is c-embedded, then the projective limit is c-embedded (Proposition 20).

Let us now state Schroder's description of c-embedded spaces:

Theorem 34

Any c-embedded space X is the projective limit of the family $\{L_{\alpha}\}_{\alpha\in M}$ of all those locally compact c-embedded spaces for which there are continous inclusions

$$X \xrightarrow{i_{\alpha}} L_{\alpha} \xrightarrow{j_{\alpha}} \beta X_s .$$

Proof:

The proof we present here is a simplified version (due to H.P.Butzmann) of Schroder's original proof. First we show that $\{L_{\alpha}\}_{\alpha\in M}$ is directed. For any two members L_{α} and L_{β} we form the intersection of the underlying sets and endow it with the initial convergence structure induced by the inclusions into L_{α} and L_{β}. The space L_{γ} obtained in this way is obviously a locally compact c-embedded space. The inclusions

$$X \xrightarrow{i_{\gamma}} L_{\gamma} \xrightarrow{j_{\gamma}} \beta X_s$$

are continuos. Hence $\{L_{\alpha}\}_{\alpha\in M}$ forms a projective system. For any compact subset $K \subset \beta X_s \setminus X_s$ the space $\beta X_s \setminus K$ belongs to $\{L_{\alpha}\}_{\alpha\in M}$. Thus the canonical map from X into $\operatorname{proj}_{\alpha\in M} L_{\alpha}$ is a continuous bijection. Let us identify the underlying sets. The canonical map j from $\operatorname{proj} L_{\alpha}$ into X_s is continuous too. To visualize the situation we use the diagram

$$X \xrightarrow{id} \operatorname{proj}_{\alpha\in M} L_{\alpha} \xrightarrow{j} X_s$$

We immediately deduce that

$$C(X) = C(\operatorname{proj}_{\alpha\in M} L_\alpha).$$

To prove that id is a homeomorphism it is sufficient to verify the identity of $C_c(X)$ and $C_c(\operatorname{proj}_{\alpha\in M} L_\alpha)$. Let therefore Ψ and Θ be filters convergent to $p\in \operatorname{proj}_{\alpha\in M} L_\alpha$ and to zero in $C_c(X)$ respectively. We will show that to any positive real number ε there are sets $P \in \Psi$ and $T \in \Theta$ related via

$$\omega(T \times P) \subset [-\varepsilon,\varepsilon].$$

Let ε be fixed. In any convergent filter Φ in X there is a set P_Φ for which

$$\omega(T_\Phi \times P_\Phi) \subset [-\varepsilon,\varepsilon]$$

for some set $T_\Phi \in \Theta$. W.l.o.g. we may assume that P_Φ is closed in X_s. We form $cl_{\beta X_s} P_\Phi$ and regard it as a subspace of βX_s. We turn $\bigcup cl_{\beta X_s} P_\Phi$ into a locally compact c-embedded space L by introducing the final convergence structure. Evidently the inclusions

$$X \xrightarrow{\ i\ } L \xrightarrow{\ j\ } \beta X_s$$

are continuous. The filter $i(\Psi)$ converges to p in L, and therefore contains, say, $\bigcup_{i=1}^{n} cl_{\beta X} P_{\Phi_i}$ for Φ_i convergent in X.

Since the members in $\{P_{\Phi_i} \mid i=1,\ldots,n\}$ are closed in X_s, the set

$$\bigcup_{i=1}^{n} P_{\Phi_i} = (\bigcup_{i=1}^{n} cl_{\beta X_s} P_\Phi) \cap X$$

belongs to Ψ. Since in addition

$$\omega(\bigcap_{i=1}^{n} T_{\Phi_i} \times \bigcup_{i=1}^{n} P_{\Phi_i}) \subset [-\varepsilon,\varepsilon]$$

holds, the proof is complete.

In connection with theorem 33, theorem 34 shows how c-embedded spaces are built up by compact topological spaces.

3.6 Continuous convergence and compact convergence. $C_I(X)$ for c-embedded X.

Let X be a c-embedded space. Moreover denote the algebra C(X) equipped with the topology of compact convergence by $C_{co}(X)$. How are $C_c(X)$ and $C_{co}(X)$ related to each other? Obviously $id: C_c(X) \longrightarrow C_{co}(X)$ is continuous. But how much finer is the continuous convergence structure than the topology of compact convergence? More precisely: Are there other locally convex vector space topologies between the continuous convergence structure and the topology of compact convergence? To find out we turn C(X) into an inductive limit called $C_I(X)$ again, for which $id : C_I(X) \longrightarrow C_c(X)$ is continuous and investigate the relationship between $C_I(X)$ and $C_{co}(X)$. We do this because $C_I(X)$, even in this generality, will be much easier to handle (with respect to the problem we posed) than $C_c(X)$.

The representation of X as $\operatorname{proj}_{\iota \in M} L_\iota$ introduced in the last section allows us to generalize the convergence structure I defined on C(X) for a completely regular topological space:

For $\iota \in M$ we consider the locally compact c-embedded convergence space L_ι. To any $p \in X$ and any filter Φ converging to p let us associate a set $F_\Phi \in \Phi$.

Since $i_\iota(\Phi) \longrightarrow p \in L_\iota$, there is a compact set $K' \in i_\iota(\Phi)$. Without loss of generality we may assume that $F_\Phi \subset K'$. Now we form in K' the closure $\bar{F}_\Phi$ of F_Φ. Let us denote the collection of compact topological spaces $\{ \bar{F}_\Phi | \Phi \longrightarrow p$ and $p \in X \}$ by Σ.

The set L given by $\bigcup_{K \in \Sigma} K$, endowed with the final convergence structure defined by the family of inclusions from all $K \in \Sigma$ into L, is a locally compact c-embedded space, containing X as a dense subset. Moreover there is a continuous inclusion $j_\iota : L \longrightarrow L_\iota$. Thus we can represent X as the projective limit $\operatorname{proj}_{\iota \in M'} L_\iota$, where $M' \subset M$ consists of all those members of M which contain X as a dense subset.

For any $\iota \in M'$ the inclusion $i_\iota : X \longrightarrow L_\iota$ induces an injection $i_\iota^* : C(L_\iota) \longrightarrow C(X)$. Let us identify each $g \in C(L_\iota)$ with $i_\iota^*(g)$. Hence $C(X) = \bigcup_{\iota \in M'} C(L_\iota)$. For any $\iota \in M'$ the continuous convergence structure on $C(L_\iota)$ is the topology of compact convergence.

The inductive limit of the family $\{ C_c(L_\iota) \}_{\iota \in M'}$ is denoted by $C_I(X)$, the convergence structure on $C_I(X)$ by I. We leave it to the reader to verify that if X is a completely regular topological space then $C_I(X)$, as just introduced, coincides with $C_I(X)$ introduced in section 2.1.

Evidently $id : C_I(X) \longrightarrow C_c(X)$ is continuous. Since $id : C_I(X_s) \longrightarrow C_I(X)$ is continuous and since every full ideal in $C_I(X)$ is closed, one has the following:

<u>Theorem 35</u>

Let X be a c-embedded convergence space. For any ideal let $N_X(J)$ be the collection of all points at which every function in J vanishes. An ideal $J \subset C_I(X)$ is closed iff $J = I(N_X(J))$, the set of all functions in $C(X)$ vanishing on $N_X(J)$. Hence a maximal ideal is closed iff it consists of all functions in $C(X)$ vanishing on a single point.

Hence for a c-embedded space X the convergence algebras $C_I(X)$, $C_c(X)$, $C_I(X_s)$, $C_c(X_s)$, $C_{co}(X)$ and thus $C_{co}(X_s)$ have all the same closed ideals.

Remark 36

Let us point out at this stage that the analogues of theorems 6,8,13 corollary 9 and proposition 14 hold for $C_I(X)$ in case X is a c-embedded convergence space. This can be obtained by modifications of the corresponding proofs. For any c-embedded convergence space X the convergence algebra $C_I(X)$ is topological iff X is locally compact.

To go back to the initial question let us take the family $\{s_\tau\}_{\tau \in N}$ (where N is an index set) of all seminorms continuous on $C_\Lambda(X)$ for some convergence structure Λ. The initial convergence structure on $C(X)$ induced by this family is a locally convex topology T. We call T the locally convex topology associated to Λ.

Our main result of this section reads now as follows:

Theorem 37

For any c-embedded convergence space X the locally convex topology on $C(X)$ associated with I is the topology of compact convergence.

Proof:

We repeat the arguments of the proof of theorem 3, p.440 in [Bi, Fe 1].

Let $s : C_I(X) \longrightarrow \mathbb{R}$ be a continuous seminorm. We construct a seminorm $\tilde{s}$ on $C(X)$ which majorizes s and is more convenient to work with. This seminorm is defined by

$$\tilde{s}(f) = \sup\{ s(g) \mid g \in C(X) \text{ and } |g| \leq |f|\}.$$

To demonstrate the continuity of $\tilde{s}$ we consider for each $\iota \in M'$ the restrictions

$$s_\iota = s|C(L_\iota) \text{ and } \tilde{s}|C(L_\iota).$$

The latter function indeed is nothing else but

$$\widetilde{s}_\iota : C(L_\iota) \longrightarrow \mathbb{R},$$

given by

$$\widetilde{s}_\iota(f) = \sup\{s(g) \mid g \in C(L_\iota) \text{ and } |g| \leq |f|\}.$$

Since s is continuous for each $\iota \in M'$ we find a compact subset $K_\iota \subset L_\iota$ and a positive real number r_ι such that

$$s_\iota \leq r_\iota \cdot \bar{s}_\iota,$$

where $\bar{s}_\iota$ assumes on each $f \in C(L_\iota)$ the value $\sup_{p \in K_\iota} |f(p)|$. Evidently

$$\widetilde{s}_\iota \leq r_\iota \cdot \bar{s}_\iota,$$

from which the continuity

$$\widetilde{s}_\iota : C_{co}(L_\iota) \longrightarrow \mathbb{R}$$

is immediate. Hence

$$\tilde{s} : C_I(X) \longrightarrow \mathbb{R}$$

is continuous too.

Next we are going to use the following properties of $\tilde{s}$:

$$\tilde{s}(f) = \tilde{s}(|f|) \quad \text{for each } f \in C(X)$$

$$\tilde{s}(f) \leq \tilde{s}(g) \quad \text{for each } f, g \in C(X) \text{ with } |f| \leq |g|,$$

which the reader is entitled to verify.

Consider now the kernel P in $C(X)$ of $\tilde{s}$. We contend that P is an ideal. To prove it consider with $f \in C(X)$ the function

$$((-\underline{n} \vee f) \wedge \underline{n}),$$

where $n \in \mathbb{N}$. Now

$$\tilde{s}(g \cdot ((-\underline{n} \vee f) \wedge \underline{n})) \leq \tilde{s}(g \cdot \underline{n}) = n \cdot \tilde{s}(g),$$

and hence

$$g \cdot ((-\underline{n} \vee f) \wedge \underline{n}) \in P \quad \text{if} \quad g \in P.$$

The Fréchet-filter of the sequence

$$\{g \cdot ((-\underline{n} \vee f) \wedge \underline{n})\}_{n \in \mathbb{N}}$$

converges in $C_I(X)$. to $g \cdot f$. Hence $g \cdot f \in P$ provided that $g \in P$. On the other hand P is an $\mathbb{R}$-algebra; thus $P \subset C_I(X)$ is a closed ideal and consists of all functions in $C(X)$ vanishing on the (non-empty) set $N_X(P) \subset X$ (theorem 35).

We will show that $N_X(P)$ is compact. Clearly $P \subset C(X)$ is the union over $\iota \in M'$ of the kernels $P_\iota \subset C(L_\iota)$ of $\widetilde{s_\iota}$. Hence

$$N_X(P) = \bigcap_{\iota \in M'} N_{L_\iota}(P_\iota),$$

Moreover $N_X(P)$ is the projective limit of $\{N_{L_\iota}(P_\iota)\}_{\iota \in M'}$. The set $N_{L_\iota}(P_\iota)$ is contained in the null set $N_{L_\iota}(\hat{P}_\iota)$ of the kernel $\hat{P}_\iota$ of $\bar{s}_\iota : C(L_\iota) \longrightarrow \mathbb{R}$. But $N_{L_\iota}(\hat{P}_\iota) = K_\iota$ is compact. Therefore $N_{L_\iota}(P_\iota)$ is compact for each $\iota \in M'$ and hence $N_X(P)$ is compact.

Next we will show that a constant multiple of the continuous semi-norm

$$\bar{s} : C_I(X) \longrightarrow \mathbb{R},$$

defined by $\bar{s}(f) = \sup_{q \in N_X(P)} |f(q)|$ for each $f \in C(X)$, majorizes s. Consider $g = (-\underline{\bar{s}(f)} \vee f) \wedge \underline{\bar{s}(f)}$, where $f \in C(X)$. We have $\tilde{s}(f-g) = o$ since $f-g$ vanishes on $N_X(P)$. Furthermore

$$|\tilde{s}(f) - \tilde{s}(g)| \leq \tilde{s}(f-g)$$

and hence $\tilde{s}(f) = \tilde{s}(g)$. From the inequality $|g| \leq \underline{\bar{s}(f)}$ we conclude

$$s(f) \leq \tilde{s}(f) \leq \tilde{s}(\underline{\bar{s}(f)}) = \bar{s}(f) \cdot \tilde{s}(\underline{1}),$$

which completes the proof.

Since $id : C_c(X) \longrightarrow C_{co}(X)$ is continuous we have in addition:

Corollary 38

For any c-embedded convergence space X the locally convex topology on $C(X)$ associated with the continuous convergence structure is the topology of compact convergence.

Another proof of corollary 38, based on an integral representation of continuous linear functionals of $C_c(X)$, for a c-embedded space X, can be found in [Bu 1].

For any convergence $\mathbb{R}$-vector space E we call the $\mathbb{R}$-vector space of all continuous real-valued linear functions of E the dual space of E, and denote it by $\mathcal{L}E$. This space equipped with the continuous convergence structure, denoted by $\mathcal{L}_c E$, is called the c-dual space of E.

Corollary 39

For any c-embedded convergence space X the convergence $\mathbb{R}$-vector spaces $C_I(X)$, $C_c(X)$ and $C_{co}(X)$ have the same dual space.

Let us point out here, that $\mathcal{L}_c C_I(X) = \mathcal{L}_c C_c(X)$ (see [Do]). However, $\mathcal{L}_c C_c(X)$ and $\mathcal{L}_c C_{co}(X)$ are different (compare Appendix, theorems 89 and 90).

4. UNIVERSAL REPRESENTATIONS OF CONVERGENCE ALGEBRAS AND SOME GENERAL REMARKS ON FUNCTION SPACES

In the previous chapter we found a class of convergence spaces, the class of c-embedded spaces, in which each object X is determined by $C_c(X)$. We exhibited this class by associating to each $C_c(X)$ (where X was any convergence space) a certain convergence space, namely $Hom_c C_c(X)$, and in addition by relating X with that space via a canonical map, namely i_X.

In this chapter we study the "dual" correspondance: Let A be a convergence algebra, that is a unitary, associative, commutative $\mathbb{R}$-algebra endowed with a convergence structure for which the operations are continuous, and assume that $Hom\, A$, the set of all continuous real-valued unitary homomorphisms, is not empty. Equipping $Hom\, A$ with the continuous convergence structure we obtain $Hom_c A$. The convergence algebra A will then be related to $C_c(Hom_c A)$ via a canonical map (which is a generalization of the well-known Gelfand map [Ri]), called the universal representation and denoted by d.

The reason we do this is that much of the structure of $C_c(X)$ is expressed in the structure of the convergence subalgebras of $C_c(X)$. We expect to turn properties of a convergence subalgebra A of $C_c(X)$ into "topological" statements on $Hom_c A$ and relate them with X.

On the other hand both correspondences - the one between X and $C_c(X)$, and the one between A and $Hom_c A$ - relate to each other and yield a satisfactory theory in connection with the continuous convergence structure on function spaces. This happens even though as yet no necessary and sufficient conditions on a subalgebra $A \subset C_c(X)$ to be dense are known.

Let us briefly outline the material we present in this chapter:

After the introduction of the universal representation and some general remarks on it, we study this representation of certain topological algebras. Here we are able to describe the structure of those topological algebras which allow d to be a bicontinuous isomorphism.

Looking at the universal representation of more general convergence algebras we restrict our attention to subalgebras of $C_c(X)$.

We prepare the investigation of $A \subset C_c(X)$ by a study on *Hom* A in relation with X.

A Stone-Weierstrass-type of theorem, which we will also use later, will then allow us to describe some aspects of the relation of a convergence subalgebra $A \subset C_c(X)$ with $C_c(X)$ itself.

Concluding remarks concern the relation of $C_c(X)$ with general function spaces.

Let us agree on the following conventions:

In this chapter when we refer to an algebra A and to a homomorphism, we mean always an associative, commutative unitary $\mathbb{R}$-algebra, and a unitary $\mathbb{R}$-algebra homomorphism, respectively. Any subalgebra $A \subset C_c(X)$ is thought to be also a subspace of $C_c(X)$, which in addition is supposed to contain the constant functions.

4.1 Universal representation of a convergence algebra

Let A be a convergence algebra. We assume $Hom\, A \neq \emptyset$. We call the space $Hom_c A$ the carrier space of A. It will allow us to consider the elements of A in a certain sense as real-valued functions defined on $Hom_c A$:

Consider the homomorphism

$$d : A \longrightarrow C_c(Hom_c A)$$

defined by $d(a)(h) = h(a)$ for any $a \in A$ and any $h \in Hom_c A$. Obviously d is continuous.

Before we study questions like what conditions on A have to be imposed such that d is injective, bijective, bicontinuous etc., let us state some general properties of d and $\mathit{Hom}_c A$ themselves.

Any homomorphism

$$u : A \longrightarrow C_c(Y)$$

induces a continuous map

$$u^* : \mathit{Hom}_c C_c(Y) \longrightarrow \mathit{Hom}_c A,$$

assuming on each $k \in \mathit{Hom}_c C_c(Y)$ the value $k \circ u$.

Hence there is a continuous map

$$u' : Y \longrightarrow \mathit{Hom}_c A$$

given by $u' = u^* \circ i_Y$. This map has the property that d followed by

$$(u')^* : C_c(\mathit{Hom}_c A) \longrightarrow C_c(Y)$$

is nothing else but u. On the other hand u' is characterized by the condition $(u')^* \circ d = u$. To prove this let $g : Y \longrightarrow \mathit{Hom}_c A$ be a map for which $g^* \circ d = u$. Then passing over to carrier spaces we have

$$u^* = d^* \circ g^{**} = d^* \circ (u')^{**} .$$

The map $d^* : \mathit{Hom}_c C_c(\mathit{Hom}_c A) \longrightarrow \mathit{Hom}_c A$ is nothing else but the homeomorphism $i^{-1}_{\mathit{Hom}_c A}$. Hence it turns out that

$$g^{**} = (u')^{**}.$$

Since the diagram

$$\begin{array}{ccc}
\mathit{Hom}_c C_c(Y) & \xrightarrow{\;f^{**}\;} & \mathit{Hom}_c C_c(\mathit{Hom}_c A) \\
\uparrow{\scriptstyle i_Y} & & \downarrow{\scriptstyle i^{-1}_{\mathit{Hom}_c A}} \\
Y & \xrightarrow{\;\;f\;\;} & \mathit{Hom}_c A
\end{array}$$

where f stands either for g or u', is commutative, we deduce the identity of g and u'.

Let us summarize this as follows:

<u>Theorem 40</u>

Let A be a convergence $\mathbb{R}$-algebra for which $Hom\, A$ is not empty. To any continuous homomorphism u from A into $C_c(Y)$, for any convergence space Y, there is a uniquely determined continuous map.

$$u' : Y \longrightarrow Hom_c A,$$

such that

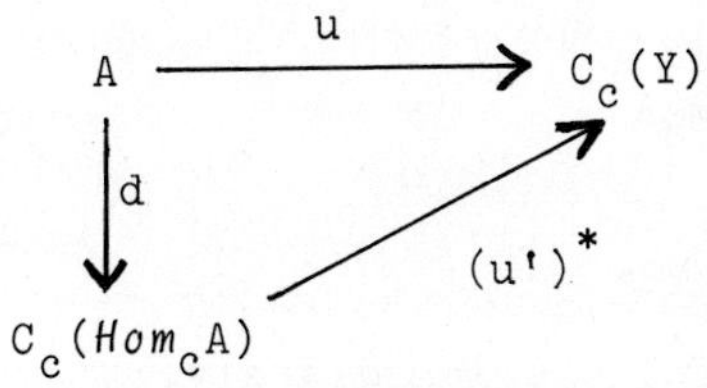

is commutative.

The next theorem affirms that $Hom_c A$ is in fact the only space (up to homeomorphism) having the universal property stated in theorem 38:

<u>Theorem 41</u>

Let X be a convergence space and

$$\bar{d} : A \longrightarrow C_c(X)$$

a continuous homomorphism.

To any continuous homomorphism u from A into $C_c(Y)$, for any convergence space Y, assume that there is a uniquely determined continuous map $u'' : Y \longrightarrow X$ with

$$(u'')^{*} \circ \bar{d} = u.$$

Then X is homeomorphic to $Hom_c A$.

<u>Proof:</u>

The uniqueness of the maps u' (in theorem 40) and u" (in theorem 41) leads to the commutative diagrams

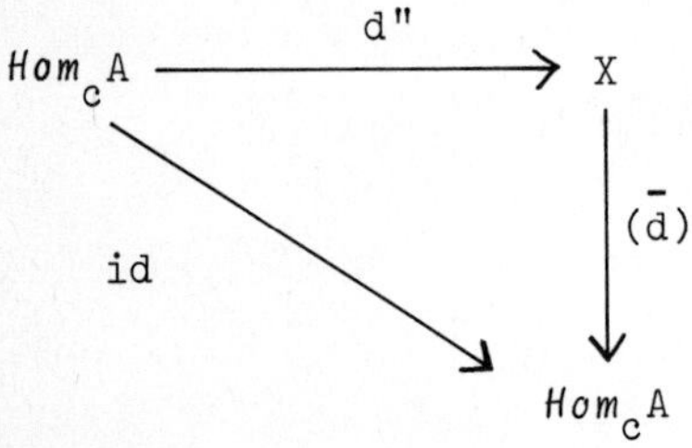

and

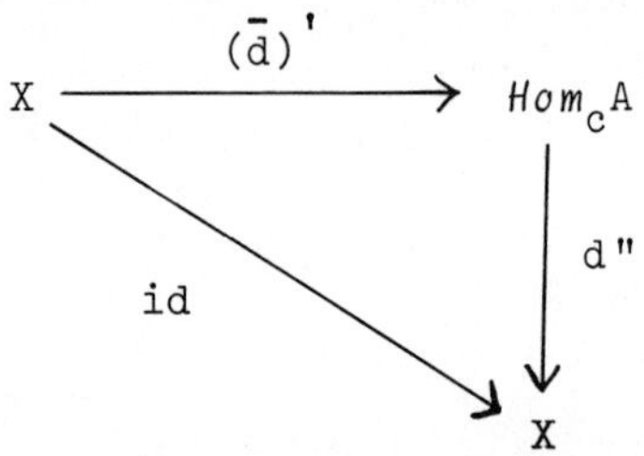

out of which the assertion follows immediately.

Because of the universal property expressed in theorem 40 we call, as in [Bi,2], the homomorphism d the <u>universal representation</u>.

We remind the reader of theorem 24, which describes the universal representation of $C_c(X)$, where X is an arbitrary convergence space.

Let us now turn our attention to topological algebras. This we do because in the case of a topological algebra the universal representati is much easier to treat than in case of general convergence algebras. In addition we observe how our theory fits into the framework of topological algebras. These preliminary investigations, however, will prepare us to recognize some key problems in the general case.

4.2 Universal representations of topological algebras

We deal in this section with a topological $\mathbb{R}$-algebra A (with a non-empty carrier space).

Corollary 29 tells us that $Hom_c A$ is a locally compact space in which every compact set is topological. Since $d(A) \subset C(Hom_c A)$ separates points in $Hom_c A$ and contains the constants, we conjecture that $d(A)$ is dense in $C_c(Hom_c A)$.

As a matter of fact we have the following generalization of the classical Stone-Weierstrass theorem, essentially presented in [Na].

Theorem 42

Let X be a c-embedded locally compact space and $A \subset C_c(X)$ a point-separating subalgebra. Then $A \subset C_c(X)$ is dense.

Proof:

Let $f \in C(X)$. We have to show that any neighborhood U of f contains some element $a \in A$. Clearly, it is enough to take U as

$$\{ g \in C(X) | s_K(f-g) < \varepsilon \},$$

where s_K is a sup-seminorm taken over a compact subset $K \subset X$, and ε is a positive real-number. Thus it remains to see that

$$(f-a)(K) \subset (-\varepsilon,\varepsilon)$$

for some $a \in A$. But this is evident by the classical Stone-Weierstrass theorem (16.4, [G,J]).

Applying theorem 42 to $d(A) \subset C_c(Hom_c A)$. we obtain the most basic theorem of the study of the universal representation of a topological $\mathbb{R}$-algebra:

Theorem 43

For any topological algebra A the universal representation

$$d : A \longrightarrow C_c(Hom_c A)$$

maps A onto a dense subalgebra.

Call an $\mathbb{R}$-algebra A real-semisimple (r.s.s.) if the intersection of the real maximal ideals (an ideal $M \subset A$ is real iff $A/M \cong \mathbb{R}$) contains the zero element only. A topological $\mathbb{R}$-algebra is said to be c-real semisimple (c-r.s.s.) if the intersection of all closed real maximal ideals contains the zero element only. For such algebras theorem 43 immediately yields:

Corollary 44

For any topological algebra A the universal representation

$$d : A \longrightarrow C_c(Hom_c A)$$

maps A isomorphically onto a dense subalgebra of $C_c(Hom_c A)$ iff A is c-r.s.s.

To conclude this section we will describe the structure of those topological algebras for which the universal representation is a bi-continuous isomorphism. Our description reads as follows:

Theorem 45

The universal representation of a topological algebra A is a bicontinuous isomorphism iff A is bicontinuously isomorphic to the projective limit of the function algebras $C_c(Y_\iota)$ of an inductive family $\{Y_\iota\}_{\iota\in M}$ of compact topological spaces. If the universal representation is a bicontinuous isomorphism then the space $\mathit{Hom}_c A$ is the inductive limit of $\{Y_\iota\}_{\iota\in M}$.

Proof:

Let d be a bicontinuous isomorphism. The space $\mathit{Hom}_c A$ is a locally compact space (Corollary 29) and each compact subspace $K \subset \mathit{Hom}_c A$ is topological. The system $\mathcal{K}$ of all compact subspaces of $\mathit{Hom}_c A$ forms, via inclusion, an inductive system. Hence $\{C(K)\}_{K\in\mathcal{K}}$ is a projective system, of which the limit can be identified with the collection of all those functions in $C(\mathit{Hom}_c A)$ which are continuous, provided that restrictions onto all compact subsets are continuous, hence with $C(\mathit{Hom}_c A)$. Since $C_c(\mathit{Hom}_c A)$ carries the topology of compact convergence it is bicontinuously isomorphic to the projective limit of $\{C_c(K)\}_{K\in\mathcal{K}}$.

Conversely assume that A is the projective limit of a family $\{C_c(Y_\alpha)\}_{\alpha\in M}$, where Y_α is compact topological for each $\alpha\in M$. This identification does not restrict the generality. For any $\alpha\in M$ there is a continuous homomorphism $r_\alpha : A \longrightarrow C_c(Y_\alpha)$. Moreover, for any choice of $\alpha,\beta \in M$ with $\alpha \leq \beta$, there is a continuous homomorphism $r^\alpha_\beta : C_c(Y_\beta) \longrightarrow C_c(Y_\alpha)$ with

$$r^\alpha_\beta \circ r_\beta = r_\alpha \, .$$

Clearly $i_{Y_\alpha} : Y_\alpha \longrightarrow \mathit{Hom}_c C_c(Y_\alpha)$ is bicontinuous. Hence $\{\mathit{Hom}_c C_c(Y_\alpha)\}_{\alpha\in M}$ forms an inductive family, where

$(r^\alpha_\beta)^* : \mathit{Hom}_c C_c(Y_\alpha) \longrightarrow \mathit{Hom}_c C_c(Y_\beta)$ is continuous for each pair

$\alpha, \beta \in M$ with $\alpha \leq \beta$. We assert that $\mathit{Hom}_c A$ is the inductive limit of the family $\{\mathit{Hom}_c C_c(Y_\alpha)\}_{\alpha \in M}$. To this end let us first show that

$$\bigcup_{\alpha \in M} r_\alpha^*(\mathit{Hom}_c C(Y_\alpha)) = \mathit{Hom}\ A.$$

The non-trivial inclusion can be verified as follows: Since A carries the initial topology induced by $\{r_\alpha\}_{\alpha \in M}$ there is to any $h \in \mathit{Hom}\ A$ an index α and homomorphism $k_\alpha \in \mathit{Hom}\ r_\alpha(A)$ with $h = k_\alpha \circ r_\alpha$. Thereby $r_\alpha(A)$ is regarded as subspace of $C_c(Y_\alpha)$. Now, k_α extends continuously to the Banach algebra $\overline{r_\alpha A}$. By theorem 43 the homomorphism $d : \overline{rA_\alpha} \longrightarrow C_c(\mathit{Hom}_c\ \overline{r_\alpha A})$ is a bicontinuous isomorphism. The inclusion $i : \overline{r_\alpha A} \longrightarrow C_c(Y_\alpha)$ induces, by theorem 40, the continuous map $i' : Y_\alpha \longrightarrow \mathit{Hom}_c\ \overline{r_\alpha A}$. Using theorem 42 one easily shows that i' is surjective. Hence k_α has an extension $\bar{h} \in \mathit{Hom}_c C_c(Y_\alpha)$. Thus $h = \bar{h} \circ r_\alpha$, and the above equation is established.

We now endow $\mathit{Hom}\ A$ with the final convergence structure induced by the family $\{r_\alpha^*\}_{\alpha \in M}$ and obtain a locally compact convergence space $(\mathit{Hom}\ A)_{1c}$ which has the universal property (with respect to the family $(r_\alpha^*)_{\alpha \in M}$) of an inductive limit. For any $a \in A$ the function $d(a)$ is therefore continuous on this space. Thus $(\mathit{Hom}\ A)_{1c}$ satisfies the criterion of c-embeddedness in theorem 33. The reader verifies easily that $d : A \longrightarrow C_c(\mathit{Hom}\ A)_{1c})$ is a bicontinuous homomorphism onto $d(A)$ regarded as a subspace of $C_c((\mathit{Hom}\ A)_{1c})$. Since the product $\prod_{\alpha \in N} A_\alpha$ is complete for any family $(A_\alpha)_{\alpha \in N}$ of complete algebras, the projective limit of such a family is, as a closed subspace, complete too. Hence $d(A) \subset C_c((\mathit{Hom}\ A)_{1c})$ is a complete subalgebra. By theorem 42 we have $d(A) = C(\mathit{Hom}\ A)_{1c})$. Since $(\mathit{Hom}\ A)_{1c}$ is c-embedded we conclude

$$\mathit{Hom}_c A = (\mathit{Hom}\ A)_{1c}.$$

4.3 *Hom* A for a subalgebra $A \subset C(X)$

The investigation of $C(X)$ where X is a convergence space requires, to a certain extent, the study of subalgebras of $C(X)$. A tool needed in this context includes the universal representations of those algebras. We prepare these studies by taking a brief look at *Hom* A, where A is a subalgebra of $C_c(X)$.

Consider the map

$$i_X^A \; : \; X \longrightarrow Hom\, A$$

defined by $i_X^A(p)\,(f) = f(p)$, for each $p \in X$ and each $f \in A$. Our first observation on *Hom* A is that in certain cases i_X^A is surjective. To formulate a criterion let us introduce A^o, the algebra of all functions in A which are bounded on X, and the notion of a lattice subalgebra $A \subset C(X)$, meaning that with any $f, g \in A$, the pointwise-defined infimum $f \wedge g$ and supremum $f \vee g$ are members of the algebra A.

Proposition 46

Let X be a convergence space and $A \subset C_c(X)$ a subalgebra for which $A^o \subset A$ is dense. Then

$$i_X^A(X) \; = \; Hom\, A.$$

Proof:

Here we follow the proof of the corresponding proposition in [Bi,Ku]. The uniform closure $\tilde{A}^o$ of A^o is a lattice subalgebra of $C(X)$ and since $h \in Hom\, A^o$ is uniformly continuous on A^o, we may extend it (continuously) on $\tilde{A}^o$, on which h is a lattice homomorphism ([G,J], § 1.6.,p.13). Assume now that $i_X^{\tilde{A}^o}(p) \neq h$ for each $p \in X$. This amounts to the existence of a function, say $f_p \in \tilde{A}^o$, with

$$h(f_p) = o \quad \text{and} \quad f_p(p) = 1$$

for each $p \in X$.

Given $f \in \tilde{A}^o$, there is to any positive real number δ a function $g \in A^o$ such that

$$||f-g|| = \sup_{q \in X} |f(q) - g(q)| < \delta.$$

On the other hand, the continuity of f_p yields a neighborhood $U_{p,\delta}$ of p on which

$$\sup_{q \in U_{p,\delta}} |f_p(q) - f_p(p)| \leqslant \delta/2.$$

Choose now $g_p \in A^o$ with $||f_p - g_p|| \leqslant \frac{\delta}{4}$. Thus we know that

$$\sup_{q \in U_{p,\delta}} |g_p(q) - 1| \leqslant \delta.$$

Moreover,

$$|h(g_p)| = |h(g_p - f_p) + h(f_p)| \leqslant |h(g_p - f_p)| \leqslant ||h|| \cdot ||f_p - g_p|| \leqslant ||f_p - g_p|| \leqslant \frac{\delta}{4}$$

For each choice of a positive real number δ with $o < \delta < \frac{1}{4}$ and $p \in X$ let us form the (non-empty) set

$$D_{\delta,p} := \{g | g \in A^o, \sup_{q \in U_{p,\delta}} |g(q) - 1| \leqslant \delta, \ h(g) \leqslant \tfrac{1}{4}\}.$$

The system $\mathcal{D} := \{D_{\delta,p} | 0 < \delta < \frac{1}{4}, \ p \in X\}$ has the finite intersection property as the following argument shows: Let $D_{\delta_1,p_1}, \ldots, D_{\delta_n,p_n}$ be members of $\mathcal{D}$. The function $f \in \tilde{A}^o$ defined by

$$f = (\bigvee_1^n f_{p_i}) \wedge \underline{1}$$

has the following properties:

a) $\sup_{q \in U_{\delta_i,p_i}} |f(q)-1| < \frac{\delta_i}{2}$ $\qquad i = 1,\dots,n.$

b) $h(f) = o.$

To $\delta = \min\{\delta_i | i=1,\dots,n\}$ there is a $g \in A^o$ with $||f-g|| \leq \frac{\delta}{2}$ which, moreover, satisfies both

$$\sup_{q \in U_{\delta_i,p_i}} |g(q)-1| \leq \delta_i \qquad i = 1,\dots,n$$

and

$$h(g) \leq \frac{1}{4}\ .$$

Hence $\bigcap_{i=1}^{n} D_{\delta_i,p_i}$ is not empty.

The filter Θ on A^o generated by $\mathcal{D}$ converges in A^o to $\underline{1}$. Since h is continuous $h(\Theta)$ converges in $\mathbb{R}$ to 1. However, $h(g) \leq 1/4$ for each g belonging to one of the sets in $\mathcal{D}$. Thus $h(\Theta)$ can not converge to 1. This contradiction requires $h|A^o$ to be a point-evaluation that means to be of the form $i_X^{A^o}(p)$ for some $p \in X$. Since A^o is dense in A and $h|A^o = i_X^{A^o}(p)$ we conclude $h = i_X^A(p)$. This completes the proof.

Since the above result is known in case X is a locally compact topological space, another way to verify the above proposition is offered:

Calling two points $p,q \in X$ equivalent if $f(q)=f(p)$ for all $f \in A$, we obtain an equivalence relation whose set of classes is denoted by X_A. For the canonical projection from X onto X_A we use the symbol Π_A. Any $f \in A$ induces a function on X_A assigning to

$\Pi_A(p)$ the value $f(p)$ for each $p \in X$. We denote this function on X_A by f_A and turn X_A into a completely regular topological space by equipping it with the initial topology defined by $\{f_A\}_{f \in A}$. Now take the preimage of A under the continuous injection $\Pi_A^* : C_I(X_A) \longrightarrow C_c(X)$ and proceed similarly as in the proof of theorem 15. For details see [Bi 7].

For any convergence space X the above proposition tells us that $Hom\, C_c^o(X) = Hom\, C_c(X)$. Since in addition $Hom_c C_c^o(X) = Hom_e C_c(X)$ we deduce that the range of the universal representation of $C_c^o(X)$ can be identified with $C_c(X)$.

A necessary and sufficient condition on $A \subset C(X)$ forcing $i_X^A(X) = Hom\, A$ is not known yet. However, for a lattice subalgebra $A \subset C(X)$, we have the following general result:

Corollary 47

Let X be a convergence space and $A \subset C_c(X)$ a lattice subalgebra. Then $i_X^A(X) = Hom\, A$.

Proof:

The sequence $((-\underline{n} \vee f) \wedge \underline{n})_{n \in \mathbb{N}}$ in A^o converges in A to f, for any $f \in A$. Thus $A^o \subset A$ is dense. Proposition 46 yields then the corollary.

The next proposition shows, that the collection of lattice subalgebras of $C(X)$ is huge. In addition it yields a method of constructing examples of lattice subalgebras of $C(X)$.

Proof:

We copy a proof of [Fe]. It is evident that the adherence of A in $C_c(X)$ is a subalgebra. To prove that $a_c(A)$ is a lattice, it suffices to show that $|f|$ belongs to $a_c(A)$ whenever f is in $a_c(A)$. This is because of the equation

$$f \vee g = \frac{1}{2}(f + g + |f-g|).$$

Let $f \in a_c(A)$, and let Θ be a filter convergent to f in $C_c(X)$ with a base in A. We denote the collection of all convergent filters on X by $\mathcal{F}$. Now for each $\Phi \in \mathcal{F}$, say $\Phi \longrightarrow p$, and each $\varepsilon > o$, there exists an $N_{\Phi,\varepsilon} \in \Phi$ and a $T_{\Phi,\varepsilon} \in \Theta$ such that

$$\omega(T_{\Phi,\varepsilon} \times N_{\Phi,\varepsilon}) \subset (f(p) - \frac{\varepsilon}{2}, f(p) + \frac{\varepsilon}{2}).$$

Define

$$D_{\Phi,\varepsilon} = \{g \in A : g(N_{\Phi,\varepsilon}) \subset (|f|(p) - \varepsilon, |f|(p) + \varepsilon).$$

We will show that $D_{\Phi,\varepsilon}$ is not empty. Indeed, we will demonstrate that for finitely many $\Phi_i \in \mathcal{F}$ and $\varepsilon_i > o$, where $i \in \{1,2...,n\}$, the set

$$\bigcap_{i=1}^{n} D_{\Phi_i,\varepsilon_i}$$

is not void. Let t be a fixed element in $\bigcap_{i=1}^{n} T_{\Phi_i,\varepsilon_i} \cap A$.

Obviously,

$$t(N_{\Phi,\varepsilon_i}) \subset (f(p_i) - \frac{\varepsilon_i}{2}, f(p_i) + \frac{\varepsilon_i}{2}),$$

where $\Phi_i \longrightarrow p_i$ for each $i \in \{1,2,...,n\}$. In particular, there exists an integer k such that

$$t(\bigcup_{i=1}^{n} N_{\Phi_i,\varepsilon_i}) \subset [-k,k].$$

Thus there exists a polynomial P with the property that for all $s \in [-1,1]$

$$|(1-s)^{1/2} - P(s)| < \frac{\varepsilon'}{2k},$$

where

$$\varepsilon' = \min\{\varepsilon_1, \varepsilon_2, \ldots, \varepsilon_n\}.$$

This means that

$$\left|\left|\frac{t}{k}\right|(p) - P(\underline{1}-(\frac{t}{k})^2)(p)\right| =$$

$$= \left|\{\underline{1}-(\underline{1}-(\frac{t}{k})^2)(p)\}^{1/2} - P(\underline{1}-(\frac{t}{k})^2)(p)\right| < \frac{\varepsilon}{2k}$$

for every $p \in \bigcup_{i=1}^{n} N_{\Phi_i,\varepsilon_i}$. Furthermore, for each $i \in \{1,2..,n\}$, we have

$$\left||f|(p_i) - kP(\underline{1}-(\frac{t}{k})^2)(p)\right| \leq$$

$$\leq \left||f|(p_i) - |t|(p)\right| + \left||t|(p) - kP(\underline{1}-(\frac{t}{k})^2)(p)\right| < \varepsilon_i,$$

for every $p \in N_{\Phi_i,\varepsilon_i}$. Hence, $kP(\underline{1}-(\frac{t}{k})^2)$ is an element of $\bigcap_{i=1}^{n} D_{\Phi_i,\varepsilon_i}$. Now the collection of sets $D_{\Phi,\varepsilon}$, for $\Phi \in \mathcal{F}$ and $\varepsilon > o$, generates a filter convergent to $|f|$ in $C_c(X)$ with a basis in A, and thus $|f| \in a_c(A)$, which completes the proof.

In connection with proposition 48 let us state the following technically very valuable proposition which is due to W.A.Feldman, from whom we copy the proof also:

Proposition 49

Let X be a convergence space. For a subset $S \subset C^{o}(X)$,

$$a_c(S) = a_c(\tilde{S}),$$

where $\tilde{S}$ denotes the uniform closure of S.

Proof:

Clearly $a_c(\tilde{S}) \supset a_c(S)$. To prove the other inclusion, assume $f \in a_c(\tilde{S})$. This means there exists a filter Θ in $C_c(X)$ such that $\Theta \longrightarrow f$ and Θ has a basis in $\tilde{S}$. Denote the collection of all convergent filters on X by $\mathcal{F}$. Now for each $\varepsilon > o$ and each $\Phi \in \mathcal{F}$, say $\Phi \longrightarrow p$, there exists an $N_{\Phi,\varepsilon} \in \Phi$ and a $T_{\Phi,\varepsilon} \in \Theta$ such that

$$\omega(T_{\Phi,\varepsilon} \times N_{\Phi,\varepsilon}) \subset [f(p) - \frac{\varepsilon}{2}, f(p) + \frac{\varepsilon}{2}] .$$

Let

$$D_{\Phi,\varepsilon} = \{g \in S : g(N_{\Phi,\varepsilon}) \subset [f(p)-\varepsilon, f(p) + \varepsilon]\},$$

and consider the collection

$$\mathcal{D} = \{D_{\Phi,\varepsilon} : \Phi \in \mathcal{F} \text{ and } \varepsilon > o\}.$$

We will show that for a finite number of elements $D_{\Phi_i,\varepsilon_i} \in \mathcal{D}$, where $i \in \{1,2,\ldots,n\}$,

$$\bigcap_{i=1}^{n} D_{\Phi_i,\varepsilon_i} \neq \emptyset.$$

First choose a function $t \in \bigcap_{i=1}^{n} T_{\Phi_i,\varepsilon_i}$. Without loss of generality we can assume $t \in \tilde{S}$ and of course

$$t(N_{\Phi_i,\varepsilon_i}) \subset [f(p_i) - \frac{\varepsilon_i}{2}, f(p_i) + \frac{\varepsilon_i}{2}],$$

where $\Phi_i \longrightarrow p_i$. Since $t \in \tilde{S}$, there exists a $g \in S$ such that $||g - t|| \leq \frac{\varepsilon_i}{2}$ for every $i \in \{1,2,\ldots,n\}$. Now for each $i \in \{1,2\ldots,n\}$, we have

$$|g(p) - f(p_i)| \leq |g(p) - t(p)| + |t(p) - f(p_i)| \leq \varepsilon_i,$$

for every $p \in N_{\Phi_i,\varepsilon_i}$. Thus $g \in \bigcap_{i=1}^{n} D_{\Phi_i,\varepsilon_i}$.

It is easy to verify that the filter generated by $\mathcal{D}$ converges to f and has a basis in S. Hence $f \in a_c(S)$, as desired.

The discussion of *Hom* A for more general types of subalgebras of $C(X)$ is based on the following result, which expresses the nature of the elements of *Hom* A:

<u>Proposition 50</u>

Let X be a completely regular topological space and $A \subset C_c(X)$ a subalgebra of $C(X)$. For any homomorphism $h \in$ *Hom* A there is a non-empty compact subset $H_h \subset \beta X$ such that each point $p \in H_h$ satisfies the equation

$$h(f) = f(p), \text{ for each } f \in A,$$

where $f(p)$ denotes the value at p of the extension of $f:X \longrightarrow \mathbb{R} \cup \{\infty\}$ onto βX.

<u>Proof:</u>

For any compact subset $K \subset \beta X \setminus X$ we define A_K to be the algebra $A \cap C(\beta X \setminus K)$ endowed with the topology of compact convergence on $\beta X \setminus K$. The identity map

$$id : \operatorname*{ind}_{K \in \mathcal{K}} A_K \longrightarrow A,$$

where $\mathcal{K}$ denotes the collection of all compact subsets of $\beta X \setminus X$, is continuous. Hence $h : \operatorname{ind}_{K \in \mathcal{K}} A_K \longrightarrow \mathbb{R}$ and thus $h_K : A_K \longrightarrow R$, which denotes the restriction of h onto A_K, is continuous for any $K \in \mathcal{K}$. Since h_K is a uniformly continuous map it has a continuous extension $\bar{h}_K$ onto $\bar{A}_K$, the closure of A_K in $C_{co}(\beta X \setminus K)$. By proposition 46 there is at least one point $q \in \beta X \setminus K$ such that

$$\bar{h}_K(f) = f(q) \quad \text{for all} \quad f \in \bar{A}_K.$$

The set H_h^K of all those points q in βX for which

$$f(q) = \bar{h}_K(f) \quad \text{for all} \quad f \in \bar{A}_K \, ,$$

is compact. For K and $K' \in \mathcal{K}$ with $K \supset K'$ we have $H_h^{K'} \supset H_h^K$. Thus

$$\bigcap_{K \in \mathcal{K}} H_h^K,$$

abbreviated by H_h, is a non-empty compact subset of βX.
Of course any point $p \in H_h$ satisfies

$$f(p) = h(f)$$

for all $f \in A$.

For any completely regular topological space X and any unitary subalgebra $A \subset C_c(X)$ the set $H_A = \bigcup_{h \in \mathit{Hom}\, A} H_h$, regarded as a subspace of βX, contains X as a dense subspace. By the above proposition the map

$$\Pi'_A : H_A \longrightarrow \mathit{Hom}_s A,$$

defined as $\Pi'_A(p)\,(f) = f(p)$ for each $p \in H_A$ and each $f \in A$, is a continuous surjection, and hence maps X onto a dense subspace. Let us point out here that by theorem 10.13 p.147 in [G,J] the map Π'_A is closed. Hence a function $f : \mathit{Hom}_s A \longrightarrow \mathbb{R}$ is continuous iff

$f \circ \Pi'_A$ is continuous. Thus $X \cong Hom_s A$ (via i_X^A) iff $H_A = X$ and A separates the points in X.

Given now an arbitrary convergence space X and a subalgebra $A \subset C_c(X)$. The canonical projection $\Pi : X \longrightarrow X_s$ induces a map

$$\Pi^{**} : Hom_s A \longrightarrow Hom_s (\Pi^*)^{-1}(A)$$

sending each $h \in Hom_s A$ into $h \circ \Pi^* | (\Pi^*)^{-1}(A)$, which is a homeomorphism onto a subspace. Here $\Pi^{*-1}(A)$ is a subspace of $C_c(X)$. Since

$$\begin{array}{ccccc} X & \xrightarrow{\Pi} & X_s & \subset & H_{(\Pi^*)^{-1}(A)} \\ {\scriptstyle i_X^A}\downarrow & & & & \downarrow{\scriptstyle \Pi'_{(\Pi^*)^{-1}(A)}} \\ Hom_s A & & \xrightarrow{\Pi^{**}} & & Hom_s (\Pi^*)^{-1}(A) \end{array}$$

commutes we end up with a sort of an estimate of $Hom\ A$ from below, in contrast with proposition 50, which states a sort of an estimate from above:

<u>Corollary 51</u>

For any convergence space X and any subalgebra $A \subset C_c(X)$ the map

$$i_X^A : X \longrightarrow Hom_s A$$

maps onto a dense subspace.

Some of the algebraic structure of A is of course expressed in the topology of $Hom_s A$, and for that reason the space $Hom_s A$ is of some interest. It seems worthwhile to us to acquaint the reader with an additional space (in connection with A and $\mathit{Hom}_s A$), which sometimes may help to visualize, by topological means, how $A \subset C(X)$ is related with X. For simplicity we assume X to be a completely regular topological space. In case $A=C(X)$, the topology of $\mathit{Hom}_s C_c(X)$ is fully determined by the compact space $\mathrm{Hom}_s C^o(X)$. Let us put this in contrast to the relations between $\mathit{Hom}_s A$ and $\widetilde{\mathrm{Hom}}_s A^o$, the space whose underlying set is the collection of all unitary, real-valued homomorphisms of A^o, continuous in the uniform norm topology and whose topology is the one of pointwise convergence. The lemma just below indicates why we use $\widetilde{\mathrm{Hom}}_s A^o$ rather than $\mathrm{Hom}_s A^o$ (the inclusion may be proper). The fact that A^o may coincide with just the constant functions in A limits the use of $\widetilde{\mathrm{Hom}}_s A^o$ for an arbitrary subalgebra $A \subset C(X)$. For simplicity we identify each $f \in A^o$ with its extension to βX.

<u>Lemma 52</u>

A necessary and sufficient condition for $h \in \mathrm{Hom}\, A^o$ to satisfy

$$h(f) = f(p),$$

for some fixed point $p \in \beta X$ and all $f \in A^o$, is the continuity of h in the uniform norm topology.

<u>Proof:</u>

The non-trivial direction is immediate by proposition 50.

Let $\tilde{A}^o$ be the uniform closure of A^o. Since $\tilde{A}^o$, endowed with the uniform norm topology, is a Banach algebra any homomorphism is continuous and the restriction map from $\mathrm{Hom}\, \tilde{A}^o$ into $\widetilde{\mathrm{Hom}}\, A^o$ is a bijection. Moreover, the spaces $\mathrm{Hom}_s \tilde{A}^o$ and $\widetilde{\mathrm{Hom}}_s A^o$ are homeomorphic. To see how X, $\mathit{Hom}_s A$, and $\widetilde{\mathrm{Hom}}_s A^o$, are related we use the canonical projection (given by lemma 52)

$$\tilde{\Pi}_{A^o} : \quad \beta X \longrightarrow \widetilde{Hom}_s A^o$$

and obtain a commutative diagram

$$\begin{array}{ccccc} X & \subset & H_A & \subset & \beta X \\ \downarrow i_X^A & & \swarrow \Pi'_A & & \downarrow \tilde{\Pi}_{A^o} \\ Hom_s A & & \xrightarrow{\quad j \quad} & & \widetilde{Hom}_s A^o \end{array} ,$$

where j is the restriction map. We immediately conclude from lemma 52:

<u>Proposition 53</u>

Let X be a completely regular topological space. For any unitary subalgebra A the space $\widetilde{Hom}_s A^o$ is compact and contains $j(Hom\ A)$ and $j(i_X^A(X))$ as dense subsets.

Clearly j is in general not even injective. In fact i_X^A may be a homeomorphism onto a subspace of $Hom_s A$ and j be a constant map. For an example of such a situation choose $X = \mathbb{R}$ and A the polynomials (over $\mathbb{R}$). If, however, enough bounded functions are available in A, as in case of a lattice subalgebra $A \subset C(X)$, then j is a homeomorphism onto a subspace. However, the relations between X and $\widetilde{Hom}_s A^o$ may even be involved in this case.

When $A = A^o$ the next proposition [Fe,1] expresses under what conditions on $\tilde{\Pi}_{A^o}$ the space X is homeomorphically embeddable into

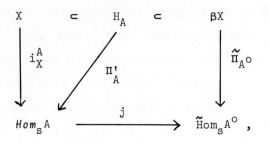

$\widetilde{\mathrm{Hom}}_s A^o$, via $j \circ i_X^A$. Aside from this, the result will provide us with the base for a technique to prove a Stone-Weierstrass type of theorem.

Proposition 54

Let X be a completely regular topological space and $A \subset C^o(X)$ a point-separating subalgebra. Then

$$\widetilde{\Pi}_A(\beta X \setminus X) = \widetilde{\mathrm{Hom}}_s A \setminus j(i_X^A(X))$$

iff $\Pi_A : X \longrightarrow X_A$ is a homeomorphism.

This proposition is a simple consequence of 10.13 p. 147 in [G.J].

4.4 A Stone-Weierstrass type of theorem

The main theorem in this section, theorem 56, is a generalization of the well-known Stone-Weierstrass theorem concerning the density of certain subalgebras in $C_c(X)$, where X is a compact topological space. The version we present here improves theorem 5 in [Bi,5] and is due to W.A.Feldman.

Before stating the theorem, let us collect some technical material. Let X be a completely regular topological space. For convenience, we again denote the extension of any $f \in C(X)$ (thought of as an $\mathbb{R} \cup \{\infty\}$-valued function) onto βX by f.

Given a unitary subalgebra $A \subset C^o(X)$, a closed set $S \subset \beta X$ is said to be A-closed if $(\widetilde{\Pi}_A)^{-1}(\widetilde{\Pi}_A(S)) = S$ holds. Since $\mathrm{Hom}_s\widetilde{A}$ and $\widetilde{\mathrm{Hom}}_s A$ are, via the restriction map, homeomorphic, the terms A - and $\widetilde{A}$-closed mean the same.

Lemma 55

Let X be a completely regular topological space. For any unitary subalgebra $A \subset C^o(X)$ the map

$$\tilde{\Pi}_A^* : C(\widetilde{Hom}_s A) \longrightarrow C(\beta X)$$

is an isomorphism onto $\tilde{A}$. Hence for any two functions $g_1, g_2 \in \tilde{A}$, there is a function $g \in \tilde{A}$ such that

$$g|S_1 = g_1|S_1 \quad \text{and} \quad g|S_2 = g_2|S_2$$

for any two disjoint A-closed sets $S_1, S_2 \subset \beta X$.

Proof:

Since $\tilde{\Pi}_A$ is a surjection $\tilde{\Pi}_A^*$ is an injection. The algebra $(\tilde{\Pi}_A^*)^{-1}(A) \subset C_c(\widetilde{Hom}_s A)$ contains the constants and separates the points. Hence it is dense in $C_c(\widetilde{Hom}_s A)$ and thus $\tilde{\Pi}_A^*$ maps onto $\tilde{A}$. The rest of the assertion follows from an extension theorem concerning continuous functions on a compact set in 3.11.c in [G,J].

If $A \subset C^o(X)$ generates the topology of X, meaning that $\Pi_A : X \longrightarrow X_A$ is a homeomorphism, then $\tilde{\Pi}_A$ maps $\beta X \setminus X$ onto $\widetilde{Hom}_s A \setminus j(i_X^A(X))$ as stated in proposition 54. This fact is basic for our technique in the forthcoming proof because it allows us to choose for any $p \in X$ and any $f \in C(X)$ an A-closed neighborhood in βX on which f stays real-valued.

Theorem 56

Let X be a completely regular topological space and $A \subset C(X)$ a subalgebra. If A^o, the algebra of all functions in A which are bounded on X, is topology generating, then A is dense in $C_c(X)$.

Proof:

The proof is based on the technique used in the proof of theorem 5 in [Bi 5] and can be found in [Fe 1]. In view of proposition 48, it is sufficient to show that $a_c(\tilde{A}^o)=C(X)$. Let f be an arbitrary element in $C(X)$. We will construct a filter Θ on $C(X)$ that converges to f in $C_c(X)$ and has a basis in $\tilde{A}^o$. For a point $p \in X$, let $g_p \in \tilde{A}^o$ be the constant function, assuming the value $f(p)$.
Define

$$V_{p,\varepsilon}=\{y \in \beta X : |f(y) - f(p)| < \varepsilon\} ,$$

for $\varepsilon > 0$. Now $V_{p,\varepsilon}$ is an open neighborhood of p in βX. This neighborhood contains an A^o-closed neighborhood of p, say $W_{p,\varepsilon}$.
Set

$$T_{p,\varepsilon} = \{g \in \tilde{A}^o : |g(y) - f(y)| < \varepsilon \quad \text{for every} \quad y \in W_{p,\varepsilon}\}.$$

Consider the collection $\mathcal{S}$ of all sets $T_{p,\varepsilon}$ for all $p \in X$ and $\varepsilon > 0$. Clearly each element $T_{p,\varepsilon} \in \mathcal{S}$ is not empty, since it contains at least the function g_p. We will show that for a finite number of elements

$$T_{p_i,\varepsilon_i} \in \mathcal{S} , \ i \in \{1,2\ldots,n\},$$

we have

$$\bigcap_{i=1}^{n} T_{p_i,\varepsilon_i} \neq \emptyset .$$

For convenience we can assume $\varepsilon_1 \leqslant \varepsilon_2 \leqslant \ldots \leqslant \varepsilon_n$. Since we know T_{p_1,ε_1} is non-empty, we assume $\bigcap_{i=1}^{m-1} T_{p_i,\varepsilon_i} \neq \emptyset$ for $m \in \{2,3,\ldots,n\}$, and prove that $\bigcap_{i=1}^{m} T_{p_i,\varepsilon_i} \neq \emptyset$. Let $L = \bigcup_{i=1}^{m-1} W_{p_i,\varepsilon_i}$. We might as well assume $W_{p_m,\varepsilon_m} \setminus L \neq \emptyset$, for otherwise our proof would be complete. Since the union of a finite number of A^o-closed sets is A^o-closed, L is an A^o - closed set. Thus $\tilde{\Pi}^{-1}_{A^o}(\tilde{\Pi}_{A^o}(y))$ is an A^o - closed set disjoint from L for every $y \in W_{p_m,\varepsilon_m} \setminus L$. Let Ω be the collection of all sets $\tilde{\Pi}^{-1}_{A^o}(\tilde{\Pi}_{A^o}(y))$ for $y \in W_{p_m,\varepsilon_m} \setminus L$. For the following calculation we will denote the elements in Ω by Greek letters.

First, we choose

$$g_1 \in \bigcap_{i=1}^{m-1} T_{p_i,\varepsilon_i} \quad \text{and} \quad g_2 \in T_{p_m,\varepsilon_m}.$$

Now for each σ and ξ in Ω, lemma 55 allows us to pick a function $g_{\sigma,\xi} \in \tilde{A}^o$ which extends both $g_1|L$ and $g_2|\sigma \cup \xi$. Let

$$M = \bigcup_{i=1}^{m} W_{p_i,\varepsilon_i}$$

(i.e., $M = L \cup W_{p_m,\varepsilon_m}$). Choose an integer k such that

$$k > \varepsilon_m + ||g_1|| + ||g_2|| \ ,$$

and set

$$f' = ((f \wedge k\underline{1}) \vee (-k\underline{1})) .$$

Clearly $f'|M = f|M$, and thus the set

$$U_\sigma^\xi = \{y \in \beta X : g_{\sigma,\xi}(y) < f'(y) + \varepsilon_m\}$$

is an open neighborhood of $\sigma \cup \xi \cup L$. For a fixed ξ, the collection $\{U_\sigma^\xi\}_{\sigma \in \Omega}$ is an open covering of the compact set M. Hence there exists a finite subset Σ_1 of Ω such that $\{U_\sigma^\xi\}_{\sigma \in \Sigma_1}$ covers M. The function

$$g_\xi = \bigwedge_{\sigma \in \Sigma_1} g_{\sigma,\xi}$$

is an element of A^o and has the property that

$$g_\xi | L = g_1 | L \,, \qquad g_\xi | \xi = g_2 | \xi \,,$$

and

$$g_\xi(y) < f'(y) + \varepsilon_m$$

for every $y \in M$. Now for each $\xi \in \Omega$, let

$$U_\xi = \{y \in \beta X : g_\xi(y) > f'(y) - \varepsilon_m\}.$$

U_ξ is an open neighborhood of $\xi \cup L$, and thus $\{U_\xi\}_{\xi\in\Omega}$ is an open covering of M. Again, there exists a finite subcovering, $\{U_\xi\}_{\xi\in\Sigma_2}$, for Σ_2 a finite subset Ω. The function

$$g = \bigvee_{\xi\in\Sigma_2} g_\xi$$

is an element of $\tilde{A}^o$ and enjoys the property that

$$g|L = g_1|L \text{ and } |g(y) - f'(y)| < \varepsilon_m$$

for every $y \in M$. Hence $g \in \bigcap_{i=1}^{m} T_{p_i,\varepsilon_i}$ as desired. It is straight forward to verify that $\mathcal{S}$ generates a filter that converges to f in $C_c(X)$ and has a basis in $\tilde{A}^o$.

Using proposition 48 we obtain the following two results:

Corollary 57

Let X be a completely regular topological space and $A \subset C^o(X)$ a subalgebra which generates the topology of X. Then the adherence of A in $C_c(X)$ is dense.

Corollary 58

Let X be a completely regular topological space and $A \subset C_c(X)$ a closed subalgebra which generates the topology of X. Then $A = C(X)$.

Since for any convergence space X the map $\Pi^*: C_c(X_s) \longrightarrow C_c(X)$ is a continuous isomorphism, theorem 56 yields:

Corollary 59

Let X be a convergence space and $A \subset C_c(X)$ a subalgebra. If A^o generates the topology of X_s, then $A \subset C_c(X)$ is dense.

As we see, corollaries 57 and 58 generalize in a similar manner to general convergence spaces. Theorem 56 is obviously a generalization of the classical Stone-Weierstrass-theorem on point-separating algebras of continuous functions on a compact space.

In this connection let us point out that in theorem 56 "topology generating" can not be weakend to "point-separating". The same is of course true of corollaries 57-59 as the following example in [Bu 3] shows: Let X be the subspace

$$([-1,0] \cap (\mathbb{R} \setminus \mathbb{Q})) \cup ([0,1] \cap \mathbb{Q})$$

of $[-1,1]$. Choose A to be the algebra of all the restrictions onto X of functions $f \in C([-1,1])$, for which $f(q) = f(-q)$ for any $q \in [0,1]$. Clearly A separates the points of X and contains the constants. However, $C(X)$ is not the adherence of A. We refer to [Bu 3] and [Sch 1] for further studies of these approximation problems.

4.5 Universal representations of subalgebras of C(X)

The symbols X and $A \subset C_c(X)$ denote an arbitrarily given convergence space and a subalgebra, respectively. We characterize first those subalgebras of $C_c(X)$ for which

$$d : A \longrightarrow C_c(Hom_c A)$$

is a bicontinuous isomorphism. If d, which is by assumption injective, is a bicontinuous bijection, then A has to be complete.

Let us suppose therefore that A is complete. The commutative diagram

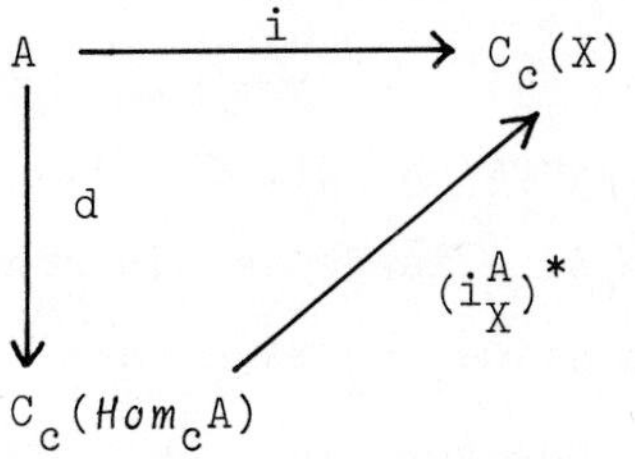

(where i denotes the inclusion map) shows that the algebra A is bicontinuously isomorphic to $d(A)$. The completeness of A does not matter here! If d is an isomorphism then $d(A)$ generates the topology of $(Hom_c A)_s$ that is $(Hom_c A)_s = Hom_s A$. On the other hand if $(Hom_c A)_s = Hom_s A$, then d has to be an isomorphism (corollary 59).

Thus we have:

Theorem 60

Let X be a convergence space and $A \subset C_c(X)$ a subalgebra. Then

$$d : A \longrightarrow C_c(Hom_c A)$$

is a homeomorphism onto $d(A)$. If A is complete then d is an isomorphism iff $(Hom_c A)_s = Hom_s A$.

To show how $C_c(Hom_c A)$ and $C_c(X)$ are related in case of more arbitrary subalgebras of $C_c(X)$ we state:

Theorem 61

Let X be a convergence space and $A \subset C_c(X)$ a subalgebra in which A^o is dense (a complete subalgebra, e.g.). Then

$$(i_X^A)^* : C_c(Hom_c A) \longrightarrow C_c(X)$$

is a continuous injection.

The proof is immediate by proposition 46.

For an arbitrary subalgebra A of $C_c(X)$, the map $(i_X^A)^*$ is an injection iff $i_X^A(X) \subset (Hom_c A)_s$ is dense. Internal conditions on A necessary and sufficient to allow $i_X(X) \subset (Hom_c A)_s$ to be dense are not known. What we may say, however, is:

Theorem 62

Let X be a convergence space and $A \subset C_c(X)$ a subalgebra. Then $(i_X^A)^* : C(Hom_s A) \longrightarrow C(X)$ is injective. If in addition A is complete, then $(i_X^A)^*$ maps onto A.

Proof:

Corollary 51 yields the first part of the theorem. The second one follows from the continuity of

$$(i_X^A)^* : C_c(Hom_s A) \longrightarrow C_c(X)$$

and corollary 58.

In conclusion of this section we characterize those $\mathbb{R}$-algebras which are isomorphic to "C(X)".

For any real-semi-simple (r.s.s.) unitary $\mathbb{R}$-algebra A, meaning that the intersection of all real maximal ideals contains o only,

$$d : A \longrightarrow C_c(\mathrm{Hom}_s A)$$

is injective. Moreover $d(A) \subset C_c(\mathrm{Hom}_s A)$ generates the topology of $\mathrm{Hom}_s A$. By equipping A with the initial convergence structure induced by d, we obviously obtain a convergence algebra. We call this convergence structure, depending on the algebra A only, the extremal continuous convergence structure, abbreviated by e.c.c.s.

Basic to our further development is:

<u>Lemma 63</u>

Let A be an r.s.s.-algebra. The space $\mathrm{Hom}_s A$ is realcompact. Hence if A is equipped with e.c.c.s. then $\mathrm{Hom}_s A = \mathit{Hom}_c A$.

<u>Proof:</u>

We have the commutative diagram

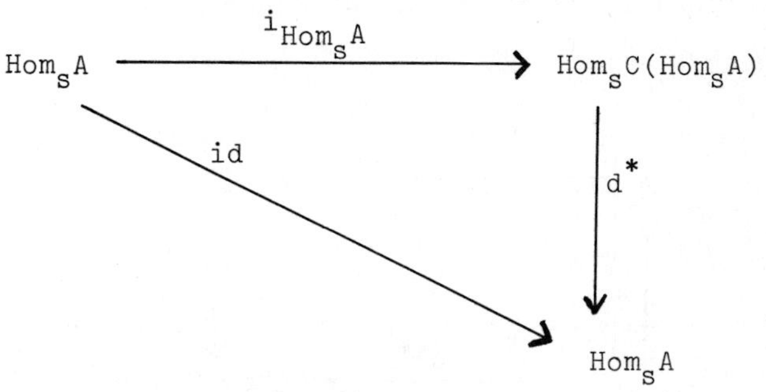

where $d : A \longrightarrow C(\mathrm{Hom}_s A)$ is defined in the obvious way. Since $i_{\mathrm{Hom}_s A}$ maps onto a dense subspace and since the above commutativity implies that $i_{\mathrm{Hom}_s A}(\mathrm{Hom}_s A)$ is closed, the first assertion made in the lemma is valued. The second one is obvious.

The following theorem characterizes those $\mathbb{R}$-algebras which "are" function algebras:

<u>Theorem 64</u>

Let A be a unitary $\mathbb{R}$-algebra. A is isomorphic to a function algebra iff

(i) A is r.s.s.

(ii) A is complete in the e.c.c.s.

<u>Proof:</u>

Assume $u : A \longrightarrow C(X)$ is an isomorphism. Then

$$u^{*} : \mathrm{Hom}_s C(X) \longrightarrow \mathrm{Hom}_s A$$

is a homeomorphism, for which

$$u^{**} : C(\mathrm{Hom}_s A) \longrightarrow C(\mathrm{Hom}_s C(X))$$

is an isomorphism. Now equip A with the e.c.c.s.
The following diagram

$$\begin{array}{ccc} A & \xrightarrow{\quad u \quad} & C(X) \\ \downarrow d & & \downarrow d_1 \\ C(\mathrm{Hom}_s A) & \xrightarrow{\quad u^{**} \quad} & C(\mathrm{Hom}_s C(X)) \end{array}$$

commutes, where $d_1(f)(h) = h(f)$ for each $f \in C(X)$ and each

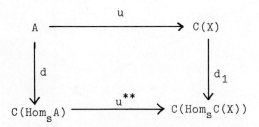

$h \in \mathrm{Hom}_s C(X)$. Since u, d_1 and u^{**} are isomorphisms, d itself is an isomorphism. Thus A is r.s.s. and complete in the e.c.c.s. If A satisfies (i) and (ii) then by lemma 63 and theorem 60

$$d : A \longrightarrow C(\mathrm{Hom}_s A)$$

is an isomorphism. For further discussions see [Bi 5].

The rest of this chapter is devoted to some aspects of function spaces which exhibit additional properties of the general behaviour of the continuous convergence structure and bring more light to $C_c(X)$ itself. Let us refer to [Bi,Ke] and [Bi 2] for papers on related topics.

4.6 Some remarks on function spaces

Let us denote by $\mathcal{L}$ the category of convergence spaces (objects: convergence spaces, morphisms: continuous maps). The full subcategory consisting of all c-embedded convergence spaces is called $\mathcal{L}_c$.

If we assign to each X in $\mathcal{L}$ the convergence algebra $C_c(X)$, belonging to the category $\mathfrak{A}$ of convergence $\mathbb{R}$-algebras (objects:convergence algebras, morphisms: continuous homomorphisms), and if we assign to each continuous map $f : X \longrightarrow Y$ the continuous homomorphism f^*, for which we choose alternatively the symbol $C_c(f)$, we then obtain a contravariant functor

$$C_c : \mathcal{L} \longrightarrow \mathfrak{A}.$$

Restricting C_c onto the category $\mathcal{L}_c$ we obtain an isomorphism onto the category $\mathfrak{A}_c$ of all convergence function algebras (objects: $C_c(X)$ where X varies over $\mathcal{L}$, morphisms: continuous homomorphisms).

The morphism set $C(Y,X)$, where $Y,X \in \mathcal{L}$, is often called a <u>function space</u>. These function spaces, endowed with the continuous convergence structure, are thus objects of $\mathcal{L}$, and are studied in various papers on various purposes, e.g. [Ba], [Bi,Ke], [Po], [C,F], etc. The purpose of this section is to illustrate how the concept of c-embeddedness is related to the behaviour of C_c on function spaces. Some of these properties are closely related to the c-embeddedness of $C_c(X,Y)$ and will thus be obtained when we derive criterions for the c-embeddedness of function spaces. The theory of universal representations will provide us with the appropriate tools.

Let any two convergence spaces X and Y be given. By theorem 25 we know that any continuous homomorphism $h : C_c(X) \longrightarrow C_c(Y)$ is induced by a continuous map, as soon as X is c-embedded. This property, in fact, is characteristic for the c-embeddedness of X:

<u>Theorem 65</u>

A convergence space X is c-embedded iff

$$C_c : C(Y,X) \longrightarrow \mathit{Hom}(C_c(X),\ C_c(Y))$$

is bijective for any convergence space Y.

<u>Proof:</u>

Let X be c-embedded. For any two maps $f,g : Y \longrightarrow X$ we consider

$$f^*,\ g^* :\quad C_c(X) \longrightarrow C_c(Y)$$

and the commutative diagram

$$\begin{array}{ccc}
Hom_c C_c(Y) & \xrightarrow{\;f^{**},\; g^{**}\;} & Hom_c C_c(X) \\
\uparrow i_Y & & \downarrow i_X^{-1} \\
Y & \xrightarrow[\;f\;,\;g\;]{} & X
\end{array}$$

Since $f^* = g^*$ implies $f^{**} = g^{**}$ we conclude the injectivity of C_c. The surjectivity of C_c is expressed in theorem 25. Conversely, for a fixed X assume that C_c is bijective for any Y. The bijectivity of

$$C_c : C(Hom_c C_c(X), X) \longrightarrow Hom(C_c(X), C_c(Hom_c C_c(X)))$$

implies that $d : C_c(X) \longrightarrow C_c(Hom_c C_c(X)))$ is induced by some continuous map $g : Hom_c C_c(X) \longrightarrow X$. Thus

$$(g \circ i_X)^* = id_{C_c(X)}.$$

The injectivity of

$$C_c : C(X,X) \longrightarrow Hom(C_c(X), C_c(X))$$

requires that $id_{C_c(X)}$ be induced by id_X. Hence $g \circ i_X = id_X$, whereupon we deduce the c-embeddedness of X.

Let us now determine conditions which guarantee the c-embeddedness of $C_c(Y,X)$. For this purpose we introduce the map

$$H : Hom_c(C_c(X), C_c(Y)) \longrightarrow C_c(Y, Hom_c C_c(X)),$$

assigning to each $u \in Hom_c(C_c(X), C_c(Y))$ the map

$$u' : Y \longrightarrow Hom_c C_c(X),$$

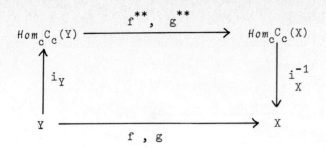

defined as in section (4.1), namely as $u^* \circ i_Y$.

The relation between C_c and $\mathcal{H}$ is visualised by the following commutative diagram:

$$(4) \qquad \begin{array}{ccc} C_c(Y,X) & \xrightarrow{\;C_c\;} & \mathit{Hom}_c(C_c(X),\, C_c(Y)) \\ \big\downarrow (i_X)_* & \swarrow \mathcal{H} & \\ C_c(Y, \mathit{Hom}_c\, C_c(X)) & & \end{array}$$

where $(i_X)_*$ sends each $g \in C(Y,X)$ into $i_X \circ g$. All the maps in this diagram are continuous.

We proceed with the solution of our problem by constructing a natural embedding of $\mathit{Hom}_c(C_c(X),\, C_c(Y))$ into $C_c(C_c(X) \times Y)$, in order to deduce the c-embeddedness of $\mathit{Hom}_c(C_c(X),\, C_c(Y))$ for any pair X,Y.

Consider $C_c(Y \times X, Z)$ and $C_c(Y, C_c(X,Z))$ for an arbitrarily given convergence space Z.

Any $f \in C(Y \times X, Z)$ restricted to $\{q\} \times X$, where $q \in Y$, defines a continuous map

$$f_q \; : \; X \longrightarrow Z,$$

assigning to each $p \in X$ the value $f_q(p) = f(q,p)$. Hence, f defines a continuous map

$$\bar{f} \; : \; Y \longrightarrow C_c(X,Z)$$

which sends $q \in Y$ into f_q.

Let us use the symbol

$$\alpha : C_c(Y \times X, Z) \longrightarrow C_c(Y, C_c(X,Z))$$

to denote the map which assigns to each $f \in C_c(Y \times X, Z)$ the map $\bar{f}$.

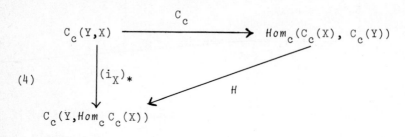

We refer to [Bi,Ke] for the proof of the following theorem:

Theorem 66

For any triple X,Y,Z the map

$$\alpha : C_c(Y \times X,Z) \longrightarrow C_c(Y,C_c(X,Z))$$

is a homeomorphism.

Now $Hom_c(C_c(X),\ C_c(Y))$ is a subspace of $C_c(C_c(X),\ C_c(Y))$. By the above theorem $C_c(C_c(X),\ C_c(Y))$ is, via α^{-1}, homeomorphic to $C_c(C_c(X) \times Y)$. Hence $Hom_c(C_c(X),\ C_c(Y))$ is homeomorphically embeddable in $C_c(C_c(X) \times Y)$. Thus by propositions 20 and 21 we have:

Lemma 67

For any choice of $X,Y \in \mathcal{L}$ the space $Hom_c(C_c(X),\ C_c(Y))$ is c-embedded.

Next we prove:

Theorem 68

For any pair of convergence spaces X,Y the map

$$H : Hom_c(C_c(X),\ C_c(Y)) \longrightarrow C_c(Y,Hom_c C_c(X))$$

is a homeomorphism.

Proof:

The continuity of $\mathcal{H}$ is based on the following diagram:

$$
\begin{array}{ccc}
Hom_c(C_c(X),\ C_c(Y)) \times Y \times C_c(X) & \xrightarrow{\ \mathcal{H} \times id_Y \times id_{C_c(X)}\ } & C_c(Y, Hom_c C_c(X)) \times Y \times C_c(X) \\
\downarrow{\scriptstyle t} & & \\
Hom_c(C_c(X),\ C_c(Y)) \times C_c(X) \times Y & & \downarrow{\scriptstyle \omega_3 \times id_{C_c(X)}} \\
\downarrow{\scriptstyle \omega_1 \times id_Y} & & \\
C_c(Y) \times Y & & Hom_c C_c(X) \times C_c(X) \\
\downarrow{\scriptstyle \omega_2} & & \downarrow{\scriptstyle \omega_4} \\
\mathbb{R} & \xrightarrow{\ id\ } & \mathbb{R}
\end{array}
$$

where t interchanges components (thus is a homeomorphism) and ω_i for $i = 1,2,3,4$ are evaluation maps. By the universal properties of continuous convergence together with the commutativity, the continuity of $\mathcal{H}$ is easily verified. To proceed further in this proof the reader is supposed to check the following:

$$(Hom^{C_c(Y)}(d) \circ C_c) \circ \mathcal{H} = id_{Hom_c(C_c(X),\ C_c(Y))},$$

where $C_c : C_c(Y, Hom_c C_c(X)) \longrightarrow Hom_c(C_c(Hom_c C_c(X),\ C_c(Y)))$

and where $Hom^{C_c(Y)}(d) : Hom_c(C_c(Hom_c C_c(X)), C_c(Y)) \longrightarrow Hom_c(C_c(X), C_c(Y))$

is defined as $Hom^{C_c(Y)}(d)(h) = h \circ d$, for any h in the domain of

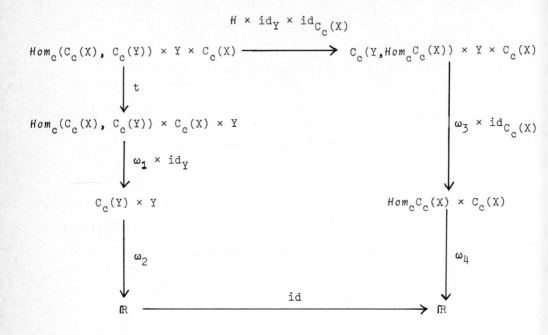

$Hom^{C_c(Y)}(d)$. The homomorphism

$$d : C_c(X) \longrightarrow C_c(Hom_c C_c(X))$$

denotes as usual the universal representation. The map

$$Hom^{C_c(Y)}(d)$$

is evidently continuous. So far we have established that H is an injective continuous map. To check that $Hom^{C_c(Y)}(d) \circ C_c$ is injective is routine. Hence

$$H^{-1} = Hom^{C_c(Y)}(d) \circ C_c ,$$

and therefore the continuity of H^{-1} is obvious.

For detailed studies of related areas we refer to [Bi 1]. Finally we point out that in the diagram (4) on page 98, the map $(i_X)_*$ is a homeomorphism iff i_X is a homeomorphism. Using the diagram (4), lemma 67, and theorem 68, we obtain the desired characterization:

Theorem 69

Let X and Y be convergence spaces. The following are equivalent

(i) X is c-embedded

(ii) $C_c(Y,X)$ is c-embedded

(iii) $C_c : C_c(Y,X) \longrightarrow Hom_c(C_c(X), C_c(X))$ is a homeomorphism.

We conclude this section by relating $C_c(Y \times X)$, which is by theorem 66 homeomorphic to $C_c(Y,C_c(X))$, with $C_c(X)$ and $C_c(Y)$, in case X and Y are completely regular topological spaces ([Fe]). The relation is based on the notion of the tensor product of $C_c(X)$ and

and $C_cY)$ in the category $\mathcal{A}_c$, consisting of the image of $C_c : \mathcal{L} \longrightarrow \mathcal{K}$.

To start let us consider $C(X) \otimes C(Y)$, the tensor product of $C(X)$ and $C(Y)$ over $\mathbb{R}$ in the category of all $\mathbb{R}$-algebras (objects: $\mathbb{R}$-algebras, morphisms: homomorphisms). We will show now that $C(X) \otimes C(Y)$ can canonically be identified with a subalgebra of $C(X \times Y)$. To this end let us consider the homomorphism

$$i_1 : C(X) \longrightarrow C(X) \otimes C(Y)$$

$$i_2 : C(Y) \longrightarrow C(X) \otimes C(Y),$$

defined by $i_1(f) = f \otimes \underline{1}$ and $i_2(q) = \underline{1} \otimes g$, for each $f \in C(X)$ and each $g \in C(Y)$.

We will relate them with the canonical projections

$$\Pi_1 : X \times Y \longrightarrow X$$

and

$$\Pi_2 : X \times Y \longrightarrow Y.$$

By the universal property of the tensor product $C(X) \otimes C(Y)$, we have a uniquely determined homomorphism

$$i : C(X) \otimes C(Y) \longrightarrow C(X \times Y)$$

for which the diagram

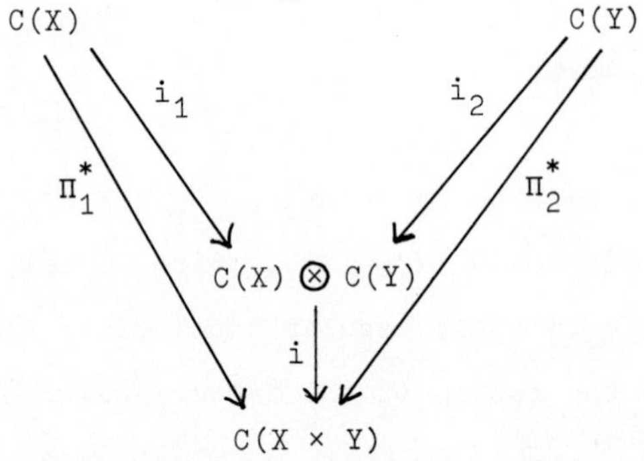

commutes.

The homomorphism i maps each generator of the form $f \otimes g \in C(X) \otimes C(Y)$ into $\Pi_1^*(f) \cdot \Pi_2^*(g)$.

It is easy to check that i is a monomorphism. For simplicity we now identify each element in $C(X) \otimes C(Y)$ with its image under i, and regard $C(X) \otimes C(Y)$ as a subspace of $C_c(X \times Y)$.

Since $C^o(X)$ and $C^o(Y)$ generate the topology of X and Y respectively the $\mathbb{R}$-algebra A^o of all bounded functions in $C(X) \otimes C(Y)$ generates the topology of $X \times Y$. Hence, by theorem 56 we have:

<u>Theorem 70</u>

For any two completely regular topological spaces X and Y the $\mathbb{R}$-algebra $C(X) \otimes C(Y)$ is dense in $C_c(X \times Y)$.

By theorem 56 the algebra A^o is dense in $C(X) \otimes C(Y)$ and thus we conclude from proposition 46 and theorem 70 that

$$i^* : \mathit{Hom}_c C_c(X \times Y) \longrightarrow \mathit{Hom}_c(C(X) \otimes C(Y))$$

is a continuous bijection. We obtain further information by investigating the universal representation of $C(X) \otimes C(Y)$.

The above diagram "dualized by taking Hom_c" yields:

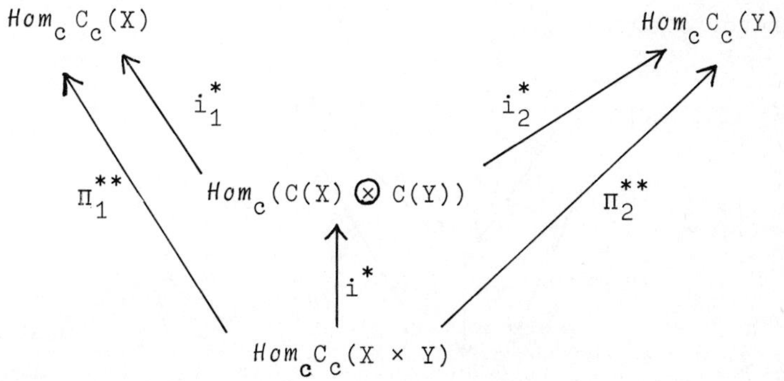

The space $Hom_c C_c(X \times Y)$ carries the initial convergence structure induced by Π_1^{**} and Π_2^{**}. Thus i^{*-1} is continuous and so i^* is a homeomorphism. We reformulate these results in:

<u>Theorem 71</u>

Let X and Y be completely regular topological spaces. Then

$$i^* : Hom_c C_c(X \times Y) \longrightarrow Hom_c(C(X) \otimes C(Y))$$

is a homeomorphism and hence

$$d : C(X) \otimes C(Y) \longrightarrow C_c(Hom_c(C(X) \otimes C(Y)))$$

maps onto a dense subspace.

Next we turn our attention to the tensor product of $C_c(X)$ and $C_c(Y)$ within the category $\mathcal{A}_c$, the image of $\mathcal{L}$ under C_c.

The <u>tensor product</u> of $C_c(X)$ and $C_c(Y)$ (for any two objects in X,Y on $\mathcal{L}_c$) in the category $\mathcal{A}_c$ is defined as an object T together with two continuous homomorphisms $h_1 : C_c(X) \longrightarrow T$ and $h_2 : C_c(Y) \longrightarrow T$ such that for any $C_c(Z)$ and any two continuous homomorphisms $k_1 : C_c(X) \longrightarrow C_c(Z)$ and $k_2 : C_c(Y) \longrightarrow C_c(Z)$ there is a unique homomorphism

$$u : T \longrightarrow C_c(Z)$$

for which the diagram

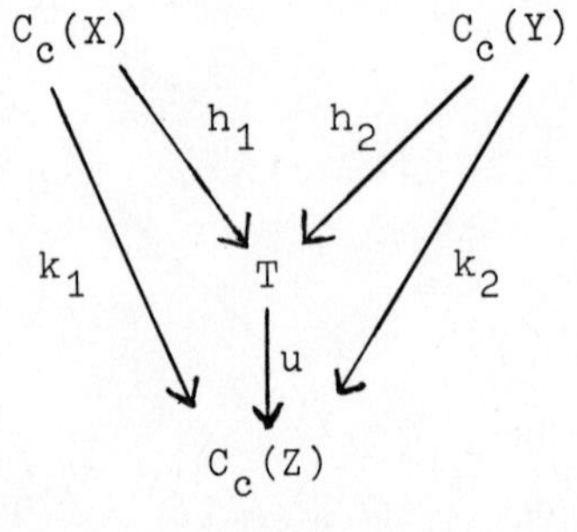

commutes.

Using theorem 71 it is easy to verify that $C_c(X \times Y)$ is (in $\mathfrak{A}_c$) the tensor product of $C_c(X)$ and $C_c(Y)$. Thus we have another relation between $C_c(X)$ and $C_c(Y)$:

<u>Theorem 72</u>

For any two completely regular topological spaces X and Y the tensor product of $C_c(X)$ and $C_c(Y)$ in $\mathfrak{A}_c$ is $C_c(X \times Y)$.

5. FUNCTIONAL ANALYTIC DESCRIPTION OF SOME TYPES OF CONVERGENCE SPACES

In this chapter some types of convergence spaces will be characterized by means of functional analytic properties of their associated convergence function algebras.

5.1 Normal topological spaces

Throughout this section X denotes a completely regular topological space.

The description of normal spaces we have in mind is based on the fact that a completely regular topological space is normal iff every closed subset is C-embedded [G,J], p.48, meaning that any continuous real-valued function on each closed set has a (continuous) extension to the whole space.

Evidently $S \subset X$ is C-embedded iff the restriction map

$$r : C(X) \longrightarrow C(S)$$

sending each $f \in C(X)$ into $f|A$ is surjective.

This version of C-embeddedness will now be reformulated in terms of completeness of some residue class algebras of $C_c(X)$.

Let $S \subset X$ be a closed non-empty subspace and $I(S)$ the ideal in $C(X)$ of all functions vanishing on S. The restriction map r allows a unique factorization over $\Pi : C(X) \longrightarrow C(X)/I(S)$, the canonical projection. This factorization $\bar{r} : C(X)/I(S) \longrightarrow C(S)$ is injective. We now endow $C(X)/I(S)$ with the final convergence structure induced by Π, obtaining a convergence algebra, denoted by $C_c(X)/I(S)$. Thus all maps in the diagram

$$\begin{array}{ccc} C_c(X) & \xrightarrow{\quad r \quad} & C_c(S) \\ & {\scriptstyle \Pi}\searrow \quad \nearrow{\scriptstyle \bar{r}} & \\ & C_c(X)/I(S) & \end{array}$$

are continuous. The following proposition shows how closely $C_c(X)$ and $C_c(S)$ are related to each other.

Proposition 73

Let X be a completely regular topological space and $S \subset X$ a closed non-empty subspace of X. The monomorphism

$$\bar{r} : C_c(X)/I(S) \longrightarrow C_c(S)$$

is a homeomorphism onto a dense subspace.

Proof:

By theorem 56 we conclude that $r(C_c(X)) \subset C_c(S)$ is dense. What is left to show is that a filter $\bar{\Theta}$ on $C(X)/I(S)$ for which $\bar{r}(\bar{\Theta})$ converges to zero in $C_c(S)$ also converges to zero in $C_c(X)/I(S)$. That is, we must construct a filter Θ on $C_c(X)$ converging to zero with the property that $\Pi(\Theta)$ is coarser than $\bar{\Theta}$. Let $\bar{\Theta}$ be a filter on $C(X)/I(S)$ with $\bar{r}(\bar{\Theta})$ convergent to zero in $C_c(S)$. Hence, for each $p \in S$ and each positive real number ε, there is a neighborhood $U_{p,\varepsilon}$ of p in X and an $F'_{p,\varepsilon} \in \bar{r}(\bar{\Theta})$ contained in $r(C(X))$ with

$$|f'(q)| \leq \varepsilon$$

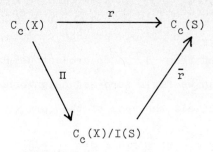

for all $f' \in F'_{p,\varepsilon}$ and all $q \in U_{p,\varepsilon} \cap S$. Without loss of generality we can assume that each $U_{p,\varepsilon}$ is a cozero-set in X. To facilitate the construction of our filter we choose inside of each $U_{p,\varepsilon}$ a zero-set neighborhood $\tilde{U}_{p,\varepsilon}$ in X of p. Furthermore, to each q in $X \setminus S$, there exists, disjoint from S, a cozero-set neighborhood V_q of q in X inside of which we fix a zero-set neighborhood $\tilde{V}_q$ of q in X. We intend to show that all the sets of the form

$$F_{p,q,\varepsilon} = \{ f \in C(X) : f|S \in F'_{p,\varepsilon} , f(\tilde{U}_{p,\varepsilon}) \subset [-2\varepsilon, 2\varepsilon] \text{ and } f(\tilde{V}_q) = \{0\}\}, \qquad (*)$$

for $p \in S$, $q \in X \setminus S$ and ε a positive real number, generate the desired filter. We first demonstrate that

$$r\left(\bigcap_{i=1}^{n} F_{p_i,q_i,\varepsilon_i} \right) \supset \bigcap_{i=1}^{n} F'_{p_i,\varepsilon_i} , \qquad (**)$$

where p_i, q_i and ε_i are as above. To this end let

$$f' \in \bigcap_{i=1}^{n} F'_{p_i,\varepsilon_i}$$

and j be a fixed integer between 1 and n. We now choose an element $f \in C(X)$ for which $r(f) = f'$ and associate to this function the sets

$$P_j = \{q \in \tilde{U}_{p_j,\varepsilon_j} : |f(q)| \geq 2 \cdot \varepsilon_j\}, \text{ and}$$

$$Q_j = \{q \in X : |f(q)| \leq \varepsilon_j\} \cup (X \setminus U_{p_j,\varepsilon_j}).$$

It is clear that $Q_j \supset S$ and, furthermore, that P_j and Q_j are disjoint zero-sets in X. Hence there is a function $h_j \in C(X)$ separating P_j and Q_j; that is,

$$h_j(q) = 0 \text{ for all } q \in P_j , \text{ and}$$
$$h_j(q) = 1 \text{ for all } q \in Q_j ,$$

Without loss of generality we may assume that $h_j(X) \subset [-1,1]$. Similarly, we pick a function $k_j \in C(X)$ with the property that

$$k_j(q) = 0 \quad \text{for all} \quad q \in \tilde{V}_y , \quad \text{and}$$
$$k_j(q) = 1 \quad \text{for all} \quad q \in X \setminus V_{y_j} ,$$

and $k_j(X) \subset [-1,1]$. The function $g = f \cdot h_1 \cdot \ldots \cdot h_n \cdot k_1 \cdots \cdot k_n$ is an element of $\bigcap_{i=1}^{n} F_{p_i,q_i,\varepsilon_i}$ and extends f'. Now the filter Θ on $C(X)$ generated by all the sets of the form (*) obviously converges to zero in $C_c(X)$. Because (**) is satisfied, $\Pi(\Theta)$ is coarser than $\bar{\Theta}$, and thus the proof is complete.

At this stage let us take a brief look at the universal representation of $C_c(X)/I(S)$. First we intend to establish a relation between $Hom_c(C_c(X)/I(S))$ and S. To do so we consider the map

$$i_X^{-1} \circ \Pi^* : Hom_c(C_c(X)/I(S)) \longrightarrow X.$$

Any function $f \in I(S)$ vanishes on $(i_X^{-1} \circ \Pi^*)(h)$ for all $h \in Hom(C_c(X)/I(S))$. Since $S \subset X$ is closed it contains $(i_X^{-1} \circ \Pi)^*(h)$.

Conversely let $p \in S$. Since $i_X(p)$ annihilates $I(S)$, it factors into

$$\overline{i_X(p)} : C(X)/I(S) \longrightarrow \mathbb{R}.$$

Since $\overline{i_X(p)} \circ \Pi = i_X(p)$, the homomorphism $\overline{i_X(p)}$ is continuous and its image under Π^* is $i_X(p)$. Thus

$$i_X^{-1} \circ \Pi^* : Hom_c(C_c(X)/I(S)) \longrightarrow S$$

is a continuous bijection.

Proposition 74

Let X be a completely regular topological space and $S \subset X$ a non-empty closed subspace. The map

$$i_X^{-1} \circ \Pi^* : Hom_c(C_c(X)/I(S)) \longrightarrow S$$

is a homeomorphism. Hence $C_c(S)$ is bicontinuously isomorphic to the universal representation of $C_c(X)/I(S)$.

Proof:

The continuity of the map $(i_X^{-1} \circ \Pi^*)^{-1}$ is clear since

$$(i_X^{-1} \circ \Pi^*)^{-1} = (\bar{r})^* \circ i_S.$$

The rest of the proposition is obvious.

The C-embeddedness of a closed subset $S \subset X$ can now be rephrased as follows:

Theorem 75

Let X be a completely regular topological space and $S \subset X$ a closed subspace. The set S is C-embedded iff $C_c(X)/I(S)$ is complete.

Proof:

If S is C-embedded then $C_c(X)/I(S)$ is bicontinuously isomorphic to $C_c(S)$ and hence is complete. The converse is an immediate consequence of proposition 73.

As a simple consequence of proposition 74 we obtain a characterization of compact subsets of X.

Proposition 76

Let X be a completely regular topological space. A closed non-empty subspace $S \subset X$ is compact iff $C_c(X)/I(S)$ is normable. If $C_c(X)/I(S)$ is normable, then it is complete.

Proof:

If S is compact then it is C-embedded ([G,J], 3.11.c, p.43) and thus $\bar{r}$ is a bicontinuous isomorphism. Since $C_c(S)$ carries the uniform norm topology, $C_c(X)/I(S)$ is normable. Conversely, if $C_c(X)/I(S)$ is normable, then by corollary 30 the space $Hom_c(C_c(X)/I(S))$ is compact. Hence S is compact. The rest is obvious.

The formulation of normality of a completely regular topological space given at the beginning of this section, combined with theorems 19 and 75, yields the desired description of normal spaces:

Theorem 77

A completely regular topological space X is normal iff $C_c(X)/J$ is complete for any closed ideal $J \subset C_c(X)$.

5.2 Separable metric spaces

It is well known that a compact topological space X is metrizable iff $C_c(X)$ is separable. By separable we mean that there is a dense countable subset. This characterization of a compact metric space X will now be extended to a description of an arbitrarily given separable metric space X in terms of $C_c(X)$. Hereby we follow [Fe].

Theorem 78

For a completely regular topological space the following are equivalent:

(i) X is metrizable and separable

(ii) $C(X)$ contains a countable topology generating subset.

(iii) $C_c(X)$ contains a countable dense topology generating subset.

Proof:

(i) ⇒ (ii): Since X allows a metric, say d, there is a map

$$j : X \longrightarrow C(X)$$

assigning to each point $p \in X$ the map $d_p : X \longrightarrow \mathbb{R}$ which sends each $q \in X$ into $d(p,q)$. The image under j of any dense subset of X generates the topology which is easily verified.

(ii) ⇒ (iii) : Let D be a countable topology generating subset of $C(X)$. W.l.o.g. we can assume $D \subset C^o(X)$. The subalgebra $A \subset C^o(X)$ generated by D generates the topology too. By theorem 56 the algebra A is dense in $C_c(X)$. We consider $\hat{D}$ consisting of all functions of the form $P(f_1 \dots, f_n)$, where $f_i \in D$ and P runs through all polynomials with rational coefficients. Clearly $\hat{D}$ is still countable. We will show now that $\hat{D} \subset C_c(X)$ is dense. Let us verify first that $\hat{D} \subset A$ is dense with respect to the uniform norm topology. Let

$$\sum_{i=1}^{n} a_i \prod_{k=1}^{m_i} f_{i_k} ,$$

for $a_i \in \mathbb{R}$ and $f_{i_k} \in D \cup \{\underline{1}\}$, be an arbitrarily given element in A. For a positive real number ε there are rationals $r_1, \dots, r_n$ such that

$$\| \sum_{i=1}^{n} a_i \prod_{k=1}^{m_i} f_{i_k} - \sum_{i=1}^{n} r_i \prod_{k=1}^{m_i} f_{i_k} \| \leq \left(\sum_{i=1}^{n} |a_i - r_i| \right) \cdot \| \prod_{k=1}^{m_i} f_{i_k} \| < \varepsilon$$

Thus the uniform closure of $\hat{D}$ contains A. By proposition 49 the set $\hat{D}$ is therefore dense in $C_c(X)$. Evidently $\hat{D}$ generates the topology of X.

(iii) = (i): Since there is a countable, topology generating subset in $C_c(X)$, the space X has a countable basis of open sets and is therefore metrizable.

We may obtain another characterization by comparing a basis of open sets of a separable metric space X with a basis of $C_c(X)$. By a basis $\mathcal{S}$ of a convergence space Y we mean a collection of subsets of Y with the following property:

Any filter $\Phi \longrightarrow p \in Y$ admits a coarser filter Φ', still convergent to p, with a filter basis consisting of members of $\mathcal{S}$. The notion of a basis of a convergence space, as just defined, generalizes the well-known notion of a basis of a topological space. A convergence space is called <u>second countable</u> if it admits a countable basis. W.A. Feldman characterized in [Fe] a separable metric space X by $C_c(X)$ in terms of the existence of a countable basis via the following general theorem:

<u>Theorem 79</u>

A c-embedded space X is second countable iff $C_c(X)$ is second countable.

<u>Proof:</u>

Let X be second countable and $\mathcal{S} = \{U_i \mid i \in \mathbb{N}\}$ be a collection of subsets of X forming a basis. Given two natural numbers, m,n and a rational, $r \in \mathbb{Q}$, we form the set $M_{m,n,r}$ defined as

$$\{t \in C(X) | t(U_m) \subset [r - \tfrac{1}{n}, r + \tfrac{1}{n}]\}.$$

The collection of all finite intersections of subsets of $C(X)$ of the form $M_{m,n,r}$ is denoted by $\mathcal{M}$. The cardinality of $\mathcal{M}$ is still countable. Let us show that it forms a basis of $C_c(X)$. Let Θ be a filter convergent to $f \in C_c(X)$. Assume that any member of Θ contains f. Let be Φ a filter convergent to, say, $p \in X$. For a given natural number n there are sets $T \in \Theta$ and $U_j \in \Phi$ belonging to $\mathcal{S}$ such that

$$\omega(T \times U_j) \subset f(p) + [-\tfrac{1}{2n}, \tfrac{1}{2n}].$$

We find a rational r such that

$$|f(p) - r| \leq \frac{1}{2n} .$$

Thus $f \in M_{j,n,r}$. Moreover, for any $g \in M_{j,n,r}$ and any $q \in U_j$ the inequality

$$|g(q) - f(p)| \leq |g(q) - r| + |r - f(p)| \leq \frac{2}{n}$$

is valid.

For any $g \in T$ and any $q \in U_j$ we therefore have

$$|g(q) - r| \leq |g(q) - f(p)| + |f(p) - r| \leq \frac{1}{n} ,$$

saying that $g \in M_{j,n,r}$.

Thus $M_{j,n,r} \supset T$ and

$$\omega(M_{j,n,r} \times U_j) \subset f(p) + [\tfrac{-2}{n} , \tfrac{2}{n}].$$

We conclude therefore that a subcollection of $\mathcal{M}$ generates a filter Θ' on $C_c(X)$ which converges to $f \in C(X)$ and which is coarser than Θ. Clearly Θ' has a basis consisting of sets in $\mathcal{M}$.

Conversely, assume that $C_c(X)$ is second countable. Any subspace of a second countable space is second countable. By the above, X itself, as a space homeomorphic to a subspace of $C_c(C_c(X))$, has to be second countable.

Any separable metric space is second countable. On the other hand any regular, second countable T_1-topological space is separable and metric ([K], p.125). In conclusion we deduce from theorem 79 a slight improvement of Feldman's description of separable metric spaces:

<u>Theorem 80</u>

A c-embedded topological space X is separable and metric iff $C_c(X)$ is second countable.

Using theorem 72 we immediately deduce:

<u>Corollary 81</u>

Let X be a c-embedded topological space, $C_c(X)$ is a separable metric space iff X is a separable metric, locally compact space.

5.3 <u>Lindelöf spaces</u>

The notion of a Lindelöf convergence space X is based on the notion of a basic subcovering of a covering system $\mathcal{S}$ of X.

<u>A basic subcovering</u> of a covering system $\mathcal{S}$ is a subfamily $\mathcal{S}'$ of $\mathcal{S}$ such that to every convergent filter Φ in X there is a finite number of elements in $\mathcal{S}'$, say $S_1,\dots,S_n$ such that $\bigcup_{i=1}^{n} S_1 \in \Phi$. According to W.A. Feldman we call a convergence space <u>Lindelöf</u> if every covering system has a countable basic subcovering. The reader can verify that a topological space is Lindelöf in the sense of Feldman precisely if every open covering allows a countable refinement.

For the characterization we intend to present let us introduce in addition the notion of <u>first countability</u>. A convergence space Y is <u>first countable</u> if every convergent filter Φ allows a coarser filter Φ' which has a countable basis and which is convergent to the same points as Φ is.

<u>Theorem 82</u>

A c-embedded convergence space X is Lindelöf iff $C_c(X)$ is first countable.

<u>Proof:</u>

Assume X is Lindelöf. Let us denote by $\mathcal{F}$ the collection of all convergent filters in X. We fix an arbitrary filter Θ convergent to $\underline{o} \in C_c(X)$. For any $n \in \mathbb{N}$ and any $\Phi \in \mathcal{F}$ there are $T_{n,\Phi} \in \Theta$ and $M_{n,\Phi} \in \Phi$ such that

$$\omega(T_{n,\Phi} \times M_{n,\Phi}) \subset [-\frac{1}{n}, \frac{1}{n}].$$

For every fixed $n \in \mathbb{N}$ the collection

$$\{M_{n,\Phi} | \Phi \in \mathcal{F}\}$$

is a covering system which admits a countable basic subcovering $\mathcal{S}$ whose sets are $M_{i,n}$ where $i=1,\dots$ Let $T_{i,n}$ be the element in $\mathcal{S}$ corresponding to $M_{i,n}$ as above. Thus we have

$$\omega(T_{i,n} \times M_{i,n}) \subset [-\frac{1}{n}, \frac{1}{n}].$$

The system $\{T_{i,n} | i,n=1,\dots\}$ generates on $C(X)$ a filter Θ' coarser than Θ and in addition has a countable basis. It remains to prove that $\Theta' \longrightarrow o \in C_c(X)$. Let $n \in \mathbb{N}$ and a filter $\Phi \in \mathcal{F}$ be given.

There are sets, say $M_{i,n} \in \mathcal{S}$, $i=1,\ldots,k$, such that their union belongs to Φ. For the corresponding sets $T_{1,n},\ldots,T_{k,n}$ we have

$$\omega(\bigcap_{i=1}^{k} T_{i,n} \times \bigcup_{i=1}^{k} M_{i,n}) \subset [-\frac{1}{n}, \frac{1}{n}],$$

which proves $\Theta' \longrightarrow o \in C_c(X)$.

To prove the converse let us suppose that $C_c(X)$ is first countable. Given a covering system $\mathcal{S}$ whose sets are denoted by S_α, where α runs through an index set M, we will prove that $\mathcal{S}$ admits a countable basic subcovering. Without loss of generality S_α may assumed to be closed in X_s for each $\alpha \in M$.

For each $S_\alpha \in \mathcal{S}$ the collection T_α defined as

$$T_\alpha = \{f \in C(X) | f(S_\alpha) = \{o\}\}$$

generates a filter which obviously converges to $\underline{o} \in C(X)$. By assumption there exists a coarser filter Θ' with a countable basis $\mathcal{D}$ which converges to $\underline{o}$. Let us denote the elements in $\mathcal{D}$ by $D_1, D_2, \ldots$ For any point $p \in X$ and a filter Φ convergent to p, there are sets $L_\Phi \in \Phi$ and $D_n \in \mathcal{D}$ with

$$\omega(D_n \times L_\Phi) \subset [-1,1].$$

For each $n \in \mathbb{N}$ we form the union R_n of all those L_Φ which correspond to D_n by the above inclusion. The collection $\{R_n | n \in \mathbb{N}\}$ forms a covering system for X, as it is easy to see. For a given $n \in \mathbb{N}$ there is a finite subset $M_n \subset \mathbb{N}$ such that

$$D_n \supset \bigcap_{\alpha \in M_n} T_\alpha .$$

We assert that $R_n \subset \bigcup_{\alpha \in M_n} S_\alpha$. Assume to the contrary that there is a point $q \in R_n \setminus \bigcup_{\alpha \in M_n} S_\alpha$. Since $\bigcup_{\alpha \in M_n} S_\alpha$ is closed in X_s there exists a function $f \in C(X)$ with $f(q)=2$ and $f(\bigcup_{\alpha \in M_n} S_\alpha)=\{o\}$. Clearly $f \in \bigcap_{\alpha \in M_n} T_\alpha$ but $f \notin D_n$. This contradicts $D_n \supset \bigcap_{\alpha \in M_n} T_\alpha$. Thus $\{S_\alpha | \alpha \in M_n$ and $n \in \mathbb{N}\}$ forms a countable basic subcovering of $\mathcal{S}$.

Next we demonstrate that not every Lindelöf convergence space is topological. To do so we choose a non-locally compact second countable completely regular topological space X. Then $C_c(X)$ is a non-topological second countable space. Combining theorems 79 and 82 we deduce that $C_c(X)$ is a Lindelöf space.

We will conclude this section by describing topological Lindelöf spaces. The proposition preparing a functional analytic reformulation of the Lindelöf property reads as follows:

Proposition 83

A completely regular topological space X is Lindelöf iff every compact set in $\beta X \setminus X$ is contained in a zero-set (of βX) in $\beta X \setminus X$.

Proof:

Assume that X is Lindelöf. Take a compact set $K \subset \beta X \setminus X$. For any $p \in X$ we choose in βX a cozero-set neighborhood U_p not meeting K. Since a countable number of these neighborhoods cover X, the set K is contained in an intersection of countably many zero-sets and hence in a zero-set Z of βZ. By construction Z does not meet X.

Conversely, suppose that any compact set $K \subset \beta X \setminus X$ is contained in some zero-set Z of βX not meeting X. In addition let $\mathcal{S}$ be an open covering of X. For each $S \in \mathcal{S}$ there is an open set S' in βX such that $S = S' \cap X$. The union of the elements of $\{S' \mid S \in \mathcal{S}\}$ constitutes an open subset of βX containing X. Let Z be a zero-set in βX which contains $\bigcap_{S \in \mathcal{S}} (\beta X \setminus S')$ and does not meet X. Clearly $\beta X \setminus Z$ is a σ-compact space, thus a Lindelöf space. Since the members of $\{S' \mid S \in \mathcal{S}\}$ cover $\beta X \setminus Z$, a countable number of them cover $\beta X \setminus Z$ and thus cover X. Hence $\mathcal{S}$ has a countable refinement.

In analogy with I on $C(X)$ (defined in §2.1) we form I', introduced in [Bi,Fe 1]: Instead of choosing the family of all compact subsets of $\beta X \setminus X$, we choose the family $\mathcal{Z}$ of those zero-sets of βX which are contained in $\beta X \setminus X$, clearly $\bigcap_{Z \in \mathcal{Z}} (\beta X \setminus Z) = \upsilon X$, the real-compactification of X. Moreover we have

$$\bigcup_{Z \in \mathcal{Z}} C(\beta X \setminus Z) = C(X).$$

We endow for every $Z \in \mathcal{Z}$ the algebra $C(\beta X \setminus Z)$ with the topology of compact convergence. The inductive limit of the family $\{C_c(\beta X \setminus Z) \mid Z \in \mathcal{Z}\}$ is the set $C(X)$ equipped with a convergence structure called I'. Evidently $I' = I$ iff every compact set $K \subset \beta X \setminus X$ is contained in a zero-set of βX not meeting X.

Thus by proposition 83 we have Kutzler's description of topological Lindelöf spaces [Ku 1]:

Theorem 84

A completely regular topological space X is Lindelöf iff

$$C_I(X) = C_{I'}(X).$$

For further discussion of I' and the original proof of theorem 84 we refer to [Ku 1]. As a concluding remark on I' let us point out that I' is topological iff υX is locally compact and σ-compact (i.e. a countable union of compact sets) [Bi,Fe 1].

5.4 $C_c(X)$ as inductive limit of topological $\mathbb{R}$-vector spaces

The next classes of spaces we are going to describe with functional analytic methods consist of completely regular topological spaces X, which require certain types of convergence structures on $C(X)$ to coincide.

For this purpose let us introduce the convergence structure of local uniform convergence on $C(X)$. A filter Θ converges locally uniformly in $C(X)$ if any point $p \in X$ admits a neighborhood U_p on which Θ converges uniformly. The algebra $C(X)$ endowed with the convergence structure of local uniform convergence is denoted by $C_{lu}(X)$. An easy verification ensures the reader that $C_{lu}(X)$ is a convergence $\mathbb{R}$-algebra and that $C_I(X) \xrightarrow{id} C_{lu}(X) \xrightarrow{id} C_c(X)$ are continuous.

We now proceed to exhibit characteristic conditions on X for $C_I(X) = C_{lu}(X)$ and for $C_{lu}(X) = C_c(X)$. First we characterize those spaces X for which $C_I(X) = C_{lu}(X)$. To do so we turn our attention to $\tilde{X}$, the collection of all those points in X which allow no compact neighborhood within υX.

Assume $\tilde{X}$ is compact and that Θ is an arbitrarily given filter on $C(X)$ which converges to $\underline{o}$ in $C_{lu}(X)$. To any point $p \in X$ there is an open neighborhood U_p in βX on which Θ converges uniformly. We may suppose that $cl_{\beta X}U_p \subset \upsilon X$ for any $p \in X \setminus \tilde{X}$. For any

$p \in X$ there is a set, say $F_p \in \Theta$, for which

$$\omega(F_p \times U_p) \subset [-1,1].$$

Finitely many of the above specified neighborhoods, say $U_{p_1},\ldots,U_{p_n}$, cover $\tilde{X}$ and satisfy

$$\omega(\bigcap_{i=1}^{n} F_{p_i} \times \bigcup_{i=1}^{n} U_{p_i}) \subset [-1,1].$$

Since $\bigcup_{p\in X\setminus\tilde{X}} U_p \subset \upsilon X$ we know that all functions in $\bigcap_{i=1}^{n} F_{p_i}$ are real-valued on $\bigcup_{i=1}^{n} U_{p_i} \cup \bigcup_{p\in X\setminus\tilde{X}} U_p$ and that Θ has a basis which is a filter convergent to $\underline{o} \in C_c(\bigcup_{i=1}^{n} U_{p_i} \cup \bigcup_{p\in X\setminus\tilde{X}} U_p)$.

Hence the compactness of $\tilde{X}$ yields $C_I(X) = C_{lu}(X)$. Let us show now that this implication is reversible. Call the set of all points in υX that admit compact neighborhoods (within υX) by $(\upsilon X)_l$. Assume that there is a family $\{U_q : q\in\tilde{X}\}$ with the property that each U_q is a closed neighborhood in βX of q and that no finite subfamily covers $\tilde{X}$. For each point $q \in X\setminus\tilde{X}$ we choose in βX a closed neighborhood U_q of q contained in $(\upsilon X)_l$. Now for each point $q \in X$, let

$$F_q = \{f \in C(X) : f(U_q) = \{0\}\}.$$

Clearly the family of all F_q for $q \in X$ generates a filter Θ convergent to zero in $C_{lu}(X)$. We claim that Θ does not converge in $C_I(X)$. Assume the contrary. That means there are points $q_1,\ldots q_n$ in X and a compact set $K \subset \beta X\setminus K$ with

$$\bigcap_{i=1}^{n} F_{q_i} \subset C(\beta X\setminus K).$$

By construction there is no finite subcollection of $\{U_q : q \in X\}$ covering $\tilde{X}$, and hence we can find a point p in the set $\tilde{X}\setminus \bigcup_{i=1}^{n} U_{q_i}$.

Furthermore, we pick in βX a closed neighborhood V of p disjoint from $K \cup \bigcup_{i=1}^{n} U_{q_i}$. Since $p \notin (\upsilon X)_1$, we know that

$$V \cap (\beta X \setminus \upsilon X) \neq \emptyset,$$

and hence there is a function

$$f \in \bigcap_{i=1}^{n} F_{q_i}$$

whose extension onto βX is not real-valued on $\beta X \setminus K$, (see [G,J],p.116). Thus f does not belong to $C(\beta X \setminus K)$.

Next let us focus on those spaces X for which $C_c(X) = C_{lu}(X)$. Here the characteristic condition on X is the countable intersection property of the neighborhood filter $\mathfrak{U}(X)$ in βX of X. This property states that any countable intersection of neighborhoods of X is a neighborhood of X. We assume that $C_c(X) = C_{lu}(X)$ and that $\mathfrak{U}(X)$ does not have the countable intersection property. Let $U_i \in \mathfrak{U}(X)$ where $i \in \mathbb{N}$ be a collection of neighborhoods whose intersection fails to be a neighborhood. Without loss of generality we can assume that $U_n \supset U_{n+1}$, $\forall n \in \mathbb{N}$. For any $p \in X$ choose in βX a closed neighborhood $U_{p,n}$ of p, contained in U_n, and put

$$F_{p,n} = \{f \in C(\beta X) \mid f(U_{p,n}) \subset [-\tfrac{1}{n}, \tfrac{1}{n}]\}.$$

The collection of all $F_{p,n}$, for $p \in X$ and $n \in \mathbb{N}$, generates a filter Θ convergent to $\underline{o} \in C_c(X)$. Our assumption requires Θ to converge locally uniformly: Since $\bigcap_{n=1}^{\infty} U_n$ is not a neighborhood of X there has to be a point, say $p \in X$, for which $\bigcap_{n=1}^{\infty} U_n$ is not a neighborhood. Thus there is in βX a neighborhood W on which Θ converges uniformly, but which is not contained in say U_n. Choose some point $q \in W \setminus U_n$. There are sets, say $F_{p_1,n_1},\dots,F_{p_k,n_k} \in \Theta$, for which

$$\bigcap_{i=1}^{k} F_{p_i,n_i}(q) \subset [-\frac{1}{n+1}, \frac{1}{n+1}].$$

We can assume that

$$n_1 \geq n_2 \geq \dots \geq n_r \geq n \geq n_{r+1} \geq \dots \geq n_k.$$

Clearly $q \notin \bigcup_{i=1}^{r} U_{p_i,n_i}$. Hence there is a function f with $|f| \leq \frac{1}{n}$, assuming the value $\frac{1}{n}$ on q and vanishing on U_{p_i,n_i} for $1 \leq i \leq r$. This function belongs to $\bigcap_{i=1}^{k} F_{p_i,n_i}$ but $f(q) \notin [-\frac{1}{n+1}, \frac{1}{n+1}]$, contradicting the above inclusion.

Finally assume that $\mathfrak{U}(X)$ has the countable intersection property. Given a filter $\Theta \longrightarrow \underline{o} \in C_c(X)$, for any $p \in X$ and any $n \in \mathbb{N}$ there is an open neighborhood $U_{p,n}$ of p in βX and say $F_{p,n} \in \Theta$ such that

$$\omega(F_{p,n} \times U_{p,n}) \subset [-\frac{1}{n}, \frac{1}{n}].$$

For each $n \in \mathbb{N}$ we define U_n to be $\bigcup_{p \in X} U_{p,n}$. By assumption $\bigcap_{n \in \mathbb{N}} U_n$ is a neighborhood of X. Let $p \in X$ be an arbitrary point and $W \subset \bigcap_{n \in \mathbb{N}} U_n$ a compact neighborhood (in βX) of p. For any positive real number ε we choose $n \in \mathbb{N}$ such that $\varepsilon > \frac{1}{n}$. Then $W \subset U_n$. Hence $W \subset \bigcup_{i=1}^{k} U_{p_i,n}$ and thus

$$\omega(\bigcap_{i=1}^{k} F_{p_i,n} \times W) \subset [-\frac{1}{n}, \frac{1}{n}],$$

showing that Θ converges uniformly on W.

Collecting the characterizations made in this section, we obtain the following improvement (due to H.P.Butzmann) of a similar theorem of [Bi,et al].

Theorem 85

For any completely regular topological space X the following hold:

(i) $C_I(X) = C_{lu}(X)$ iff $\tilde{X}$ is compact.

(ii) $C_c(X) = C_{lu}(X)$ iff the neighborhood filter in βX of X has the countable intersection property.

For another proof of these results one may consult [Ku 2].

The above theorem allows us to give precise conditions on X such that $C_c(X)$ be an inductive limit of topological $\mathbb{R}$-vector spaces. In fact the following two results of [Bi, et al] and [Fe] are easily deducible from corollary 9 and theorem 85:

Theorem 86

For a completely regular topological space X the convergence $\mathbb{R}$-algebra $C_c(X)$ is an inductive limit of topological $\mathbb{R}$-vector spaces iff $C_c(X) = C_I(X)$, that is iff $\tilde{X}$ is compact and the neighborhood filter in βX of X has the countable intersection property.

Corollary 87

For a first countable completely regular topological space X (e.g. a metric space) the convergence algebras $C_c(X)$ and $C_I(X)$ coincide iff X is locally compact.

Proof:

We repeat a proof given in [Bi, et al]. If X is locally compact, then $C_c(X) = C_I(X)$. To prove the converse assume that $p \in X$ does not admit a compact neighborhood. Since X is first countable there is a sequence $(p_i)_{i \in \mathbb{N}}$ of points in $\beta X \setminus X$ converging to p. Evidently $\bigcap_{i=1}^{\infty} \{\beta X \setminus \{p_i\}\}$ is, in βX, not a neighborhood of X. Hence $C_c(X) \neq C_I(X)$.

The reader quickly realizes that $C_c(\mathbb{Q})$, $C_{lu}(\mathbb{Q})$ and $C_I(\mathbb{Q})$ are all different.

To see that $C_c(X) = C_I(X)$ does not imply the local compactness of X, we construct a space X as follows:

Choose an uncountable set X. For any point $p \in X$ except for one, say $q \in X$, let the neighborhood filter be $\dot{p}$. We assume the neighborhood filter of q to be generated by all sets containing q and having a countable complement. Clearly X is not locally compact, however, the conditions in theorem 86 yielding $C_c(X)=C_I(X)$ are satisfied [Bi, et al].

APPENDIX. SOME FUNCTIONAL ANALYTIC ASPECTS OF $C_c(X)$

We will be concerned with two "dualities" of $C_c(X)$, the linear duality and the Pontryagin's duality. Both indicate the very special character of the continuous convergence structure Λ_c in this part of functional analysis.

1. The c-Reflexivity of $C_c(X)$

For any convergence space X and any subset S let us denote by VS the linear space spanned by $i_X(S) \subset \mathcal{L}_c C_c(X)$. The man theorem, proved in [Bu 2], via an integral representation of positive linear functionals, is the following:

Theorem 88

For any convergence space X the linear space VX is dense in $\mathcal{L}_c C_c(X)$.

Proof:

First let us verify the assertion in case X is a compact topological space. The space $\mathcal{L}_c C_c(X)$ is locally compact, by the assumption just made. In fact the unit-ball U of $\mathcal{L} C_c(X)$ is compact and carries the topology of pointwise convergence (proposition 28). Hence U is the closed convex hull of $i_X(X) \cup (-i_X(X))$ (see [Du,Sch], V 8.6). Since U is absorbant, $VX \subset \mathcal{L}_c C_c(X)$ has to be dense.

Next let X be c-embedded. Choose $u \in \mathcal{L}_c C_c(X)$ By corollary 38 the seminorm $|u|$ can be majorized by a real multiple of a sup-seminorm taken over some compact set $K \subset X$. Since the restriction map

$r : C_c(X) \longrightarrow C_c(K)$ induces a continuous injection $r^* : \mathcal{L}_c C_c(K) \longrightarrow \mathcal{L}_c C_c(X)$ and since u factors over r, it is in the closure of VK. Thus $VX \subset \mathcal{L}_c C_c(X)$ is dense.

Finally let X be an arbitrary convergence space. The surjective map $i_X : X \longrightarrow Hom_c C_c(X)$ induces a bicontinuous isomorphism between $C_c(Hom_c C_c(X))$ and $C_c(X)$ and hence between $\mathcal{L}_c C_c(X)$ and $\mathcal{L}_c C_c(Hom_c C_c(X))$. Therefore $VX \subset \mathcal{L}_c C_c(X)$ is dense.

A convergence $\mathbb{R}$-vector space E is called <u>c-reflexive</u> if the canonical mapping

$$j_E : E \longrightarrow \mathcal{L}_c \mathcal{L}_c E,$$

defined by $j_E(t)(u) = u(t)$ for all $t \in E$ and for all $u \in \mathcal{L}_c E$, is a bicontinuous isomorphism.

Let us consider $j_{C_c}(X)$ for an arbitrary convergence space X. The map $i_X : X \longrightarrow \mathcal{L}_c C_c(X)$ induces a map

$$\tilde{i}_X : \mathcal{L}_c \mathcal{L}_c C_c(X) \longrightarrow C_c(X),$$

defined by $\tilde{i}_X(k) = k \circ i_X$ for each $k \in \mathcal{L}_c \mathcal{L}_c C_c(X)$, which obviously is continuous. Moreover

$$\tilde{i}_X \circ j_{C_c(X)} = id_{C_c(X)} .$$

Thus by theorem 88 we obtain the c-reflexivity of $C_c(X)$ (see [Bu 2]):

<u>Theorem 89</u>

For any convergence space X the convergence $\mathbb{R}$-algebra $C_c(X)$ is c-reflexive.

This result allows us to treat questions on the c-reflexivity of topological $\mathbb{R}$-vector spaces.

Let E be a topological $\mathbb{R}$-vector space. Since $\mathcal{L}_c E$ is locally compact, $\mathcal{L}_c \mathcal{L}_c E$ is a locally convex topological $\mathbb{R}$-vector space. Thus E can be c-reflexive only if E is locally convex. Let us assume that this is satisfied.

Any locally convex topological $\mathbb{R}$-vector space can be embedded in a $C_c(X)$ for some locally compact space X, (see [Kö], § 20, p.107). Call the embedding e.

Use the universal property of the continuous convergence and the Hahn-Banach theorem to verify that

$$e^*: \mathcal{L}_c C_c(X) \longrightarrow \mathcal{L}_c E$$

(sending each u into $u \circ e$) is a surjection, and that

$$e^{**}: \mathcal{L}_c \mathcal{L}_c E \longrightarrow \mathcal{L}_c \mathcal{L}_c C_c(X)$$

is a bicontinuous isomorphism onto a subspace.

The commutative diagram

$$\begin{array}{ccc} E & \xrightarrow{\quad e \quad} & C_c(X) \\ \Big\downarrow j_E & & \Big\downarrow j_{C_c(X)} \\ \mathcal{L}_c \mathcal{L}_c E & \xrightarrow{\quad e^{**} \quad} & \mathcal{L}_c \mathcal{L}_c C_c(X) \end{array}$$

yields, together with theorem 89, that j_E is a homeomorphism onto a subspace.

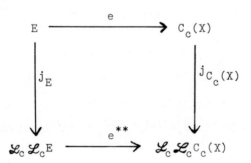

Let us show that j_E maps onto a dense subspace. To this end consider $\mathcal{L}_c E$. Clearly $\mathcal{L}_c \mathcal{L}_c E$ is a subspace of the locally convex topological space $C_c(\mathcal{L}_c E)$ and the restriction map

$$r : \mathcal{L}_c C_c(\mathcal{L}_c E) \longrightarrow \mathcal{L}_c \mathcal{L}_c \mathcal{L}_c E$$

is surjective (Hahn-Banach). Now $r \circ i_{\mathcal{L}_c E} = j_{\mathcal{L}_c E}$. Thus $r(V(\mathcal{L}_c E))$ is mapped onto a dense subspace of $\mathcal{L}_c \mathcal{L}_c \mathcal{L}_c E$ (theorem 88), which moreover has to be $j_{\mathcal{L}_c E}(\mathcal{L}_c E)$.

Since $j_E^* \circ r \circ j_{\mathcal{L}_c E} = id_{\mathcal{L}_c E}$, where j_E^* maps each $u \in \mathcal{L}_c \mathcal{L}_c \mathcal{L}_c E$ into $u \circ j_E$, we deduce that $j_{\mathcal{L}_c E}(\mathcal{L}_c E)$ is closed. Hence it is the whole space $\mathcal{L}_c \mathcal{L}_c \mathcal{L}_c E$.

This demonstrates that $\mathcal{L}_c E$ is c-reflexive. Thus E and $\mathcal{L}_c \mathcal{L}_c E$ have bicontinuously isomorphic dual spaces, which is only possible if $j_E(E) \subset \mathcal{L}_c \mathcal{L}_c E$ is dense. In conclusion let us state:

<u>Theorem 90</u>

For any locally convex topological $\mathbb{R}$-vector space E the map

$$j_E : E \longrightarrow \mathcal{L}_c \mathcal{L}_c E$$

is a homeomorphism onto a dense image. Hence a topological $\mathbb{R}$-vector space E is c-reflexive iff E is locally convex and complete.

The c-duals of topological $\mathbb{R}$-vector spaces are characterized as those locally compact convergence $\mathbb{R}$-vector spaces in which any compact subspace is topological and which possess point-separating continuous functionals.

For papers related to the c-duality we refer to [Bu 2] and [Bi,Bu,Ku]. The relation of theorem 90 to the classical results concerning the completion of locally convex $\mathbb{R}$-vector spaces can be made via an Ascoli-Arzelà theorem in [C,F].

2. On the Pontryagin reflexivity of certain convergence $\mathbb{R}$-vector spaces

Let E be a Hausdorff $\mathbb{R}$-vector space and T be the group (with the ususal topology) of all complex numbers of modulus one.

The collection of all group homomorphisms of E into T forms a group GE and if endowed with the continuous convergence structure a convergence group G_cE, then it is called the character group of E. Introducing the character group G_cG_cE of G_cE, we see that the canonical map

$$\hat{j}_E : E \longrightarrow G_cG_cE,$$

defined by $\hat{j}_E(e)(h) = h(e)$ for all $e \in E$ and all $h \in G_cE$, is continuous.

We now proceed to determine for which type of convergence $\mathbb{R}$-vector space E the map $\hat{j}_E$ is a bicontinuous isomorphism; that is to say which space E is <u>Pontryagin reflexive</u>. The methods we use here are based on the fact that $\mathbb{R}$ is a (the universal) covering of T. The covering projection $\kappa : \mathbb{R} \longrightarrow T$ sends each $r \in \mathbb{R}$ into $e^{2\pi i r}$.

For every Hausdorff convergence $\mathbb{R}$-vector space E, the convergence structure induces on any finite dimensional subspace the natural topology [Ku 3]. Now let $h \in G_cE$. The restriction of h to any finite dimensional subspace H lifts to a unique, continuous linear functional u_H of H. Thus h lifts uniquely to a linear functional u of E. To prove the continuity of u let us suppose that E is <u>balanced</u>, i.e. that with any filter Φ convergent to $o \in E$, the filter $[-1,1]\cdot\Phi$ converges too. Since $[-1,1]\cdot\Phi$ has a basis of path connected sets (defined in the obvious way), the convergence of $u([-1,1]\cdot\Phi)$ to zero is obvious. Hence u is continuous. For any $u \in \mathcal{L}E$ the map $\kappa \circ u$ is a character of E.

Hence

$$\kappa_E : \mathcal{L}E \longrightarrow GE,$$

sending each u into $\kappa \circ u$, is a group isomorphism. Since κ is a local homeomorphism and $\mathcal{L}_c E$ is path connected, κ_E is bicontinuous if both spaces are equipped with the continuous convergence structure.

By making explicit use of the theory of coverings, we have just obtained a short proof (and seen a variety of possible generalizations) of the following theorem (see [Kö] p.313 and [Bu 1]).

Theorem 91

For every balanced Hausdorff convergence $\mathbb{R}$-vector space E the covering map $\kappa : \mathbb{R} \longrightarrow T$ induces a bicontinuous group isomorphism

$$\kappa_E : \mathcal{L}_c E \longrightarrow G_c E.$$

Considering the following commutative diagram

$$\begin{array}{ccc} E & \xrightarrow{\hat{j}_E} & G_c G_c E \\ \Big\downarrow{\scriptstyle j_E} & & \Big\downarrow{\scriptstyle \kappa_E^*} \\ \mathcal{L}_c \mathcal{L}_c E & \xrightarrow{\kappa_{\mathcal{L}_c E}} & G_c \mathcal{L}_c E \end{array}$$

where $\kappa_E^*(k) = k \circ \kappa_E$ for each $k \in G_c G_c E$, we immediately deduce [Bu 1]:

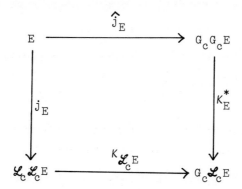

Corollary 92

For every balanced Hausdorff convergence $\mathbb{R}$-vector space E the map

$$b : \mathcal{L}_c\mathcal{L}_c E \longrightarrow G_c G_c E,$$

where $b = \kappa_E^{*-1} \circ \kappa_{\mathcal{L}_E}$, is a bicontinuous isomorphism. Thus E is Pontryagin reflexive iff it is c-reflexive.

With the results of the last section applied to $C_c(X)$, the following corollary [Bu 1] is immediate:

Corollary 93

For any convergence space X the convergence $\mathbb{R}$-algebra $C_c(X)$ is Pontryagin reflexive. The character group $G_c C_c(X)$ is the closure of $\kappa_{C_c(X)}(VX)$.

Let us add to these results the general description of those topological $\mathbb{R}$-vector spaces which are Pontryagin reflexive (see [Bi 6]).

Theorem 94

A topological $\mathbb{R}$-vector space E is Pontryagin reflexive iff E is a complete locally convex space.

The proof is easily made by combining theorem 90 and corollary 92.

The general correspondence between complete subspaces of complete locally convex spaces and the whole character groups modulo annihilators, characteristic for Pontryagin's duality theory, are valid and are easy to verify.

B I B L I O G R A P H Y

[Ba] A.Bastiani: Applications différentiables et variétés différentiables de dimension infinie. J.Analyse Math. 13 (1964), 1-114.

[Bi 1] E.Binz: Bemerkungen zu limitierten Funktionenalgebren. Math. Ann. 175 (1968), 169-184.

[Bi 2] ——: Zu den Beziehungen zwischen c-einbettbaren Limesräumen und ihren limitierten Funktionenalgebren. Math. Ann. 181 (1969), 45-52.

[Bi 3] ——: On closed ideals in convergence function algebras. Math. Ann. 182 (1969), 145-153.

[Bi 4] ——: Kompakte Limesräume und limitierte Funktionenalgebren. Comment. Math. Helv. 43 (1968), 195-203.

[Bi 5] ——: Notes on a characterization of function algebras. Math. Ann. 186 (1970), 314-326.

[Bi 6] ——: Recent results in the functional analytic investigations of convergence spaces. General topology and its relations to modern analysis and algebra III, proceedings of the third Prague topological symposium (1971), 67-72.

[Bi 7] ——: Representations of convergence algebras as algebras of real-valued functions. Convegno sugli "Anelli di funzioni continue", Rome Nov. 73
To appear in Symposia Mathematica.

——: Functional analytic methods in topology. Convegno di "Topologia insiemistica e generale", Rome March 73.
To appear in Symposia Mathematica.

[Bi,Bu,Ku] E.Binz, H.P.Butzmann and K.Kutzler: Über den c-Dual eines topologischen Vektorraumes. Math.Z. 127 (1972), 70-74.

——: Bemerkungen über eine Klasse von ℝ-Algebren-Topologien auf C(X). Arch.Math. 23 (1972), 80-82.

[Bi,Fe 1] E.Binz and W.A.Feldman: On a Marinescu structure on C(X). Comment.Math.Helv. 46 (1971), 436-450.

[Bi,Fe 2] ——: A functional analytic description of normal spaces. Canad.J.Math. 24 (1) 1972, 45-49.

[Bi,Ke] E.Binz and H.H.Keller: Funktionenräume in der Kategorie der Limesräume. Ann.Acad.Sci.Fenn.Ser. A I. 383 (1966) 1-21.

[Bi,Ku] E.Binz and K.Kutzler: Über metrische Räume und $C_c(X)$. Ann.Scuola Norm.Sup.Pisa, Vol. 26, Fasc.I (1972), 197-223.

[Bi et al] E.Binz, H.P.Butzmann, W.Feldmann, K.Kutzler and M.Schroder: On ω-admissible vector space topologies on C(X). Math.Ann. 196 (1972.

[Bou] N.Bourbaki: Elements of mathematics. General topology, Part 1, Addison-Wesley, 1966.

[Bu 1] H.P.Butzmann: Dualitäten in $C_c(X)$. Ph.D.Thesis, University of Mannheim, W.Germany (1971).

[Bu 2] ——: Über die c-Reflexivität von $C_c(X)$. Comment.Math. Helv. 47 (1972), 92-101.

[Bu 3] ——: Der Satz von Stone-Weierstrass in $C_c(X)$ und seine Anwendungen auf die Darstellungstheorie von Limesalgebren. Habilitationsschrift at the University of Mannheim (1974).

[Bu,Mü] H.P.Butzmann and B.Müller: Topological, c-embedded spaces, to appear.

[C,F] C.H.Cook and H.R.Fischer: On equicontinuity and continuous convergence. Math.Ann.159 (1965), 94-104.

[Do] K.P.Dostmann: Über die c-Reflexivität und die kompakten Teilmengen von $C_I(X)$. Ph.D.Thesis, University of Mannheim, W.Germany (1974).

[Du,Sch] N.Dunford and J.Schwartz: Linear operators, Part.I, Interscience Publishers Inc., New York.

[Fe] W.A.Feldman: Topological spaces and their associated convergence function algebras. Ph.D.Thesis, Queen's Univ., Kingston, Canada, (1971).

[Fe] ——: Axioms of countability and the algebra C(X). Pac.J.Math., Vol.47, No.1, (1973), 81-89.

——: A characterization of the topology of compact convergence on C(X). Pac.J.Math., Vol.51, No.1, (1974).

[Fi] H.R.Fischer: Limesräume, Math.Ann.137, (1959), 169-303.

A.Frölicher: Sur la transformation de Dirac d'un espace à génération compacte. Deuxième colloque d'analyse fonctionelle, Bordeaux 1973. To appear in: Lecture Notes, Springer.

A.Frölicher and W.Bucher: Calculus in vector spaces without norm. Lecture Notes in Mathematics 30 (1966), Springer Berlin-Heidelberg-New York.

A.Frölicher and H.Jarchow: Zur Dualitätstheorie kompakterzeugter und lokalkonvexer Vektorräume. Comment.Math. Helv.47, (1972), 289-310.

[G,J] L.Gillman and M.Jerison: Rings of continuous functions. Nostrand Series in Higher Mathematics, 1960.

[Ja] H.Jarchow: Marinescu Räume, Comment.Math.Helv.44, (1969), 138-163.

[Ja] ——: Duale Charakterisierungen der Schwartz-Räume. Math.Ann.196, (1972), 85-90.

M.Katětov: Convergence structures. General topology and its relations to modern analysis and algebra II. Proceedings of the second Prague topological symposium, (1966), 207-216.

[Ke] H.H.Keller: Differential calculus in locally convex spaces. Lecture Notes in Mathematics 417, Springer-Verlag Berlin-Heidelberg-New York.

[K] J.L.Kelley: General topology. Van Nostrand, Princeton 1968.

[Kö] G.Köthe: Topologische Vektorräume. Grundlehren der Mathematischen Wissenschaften Bd. 107. Springer-Verlag Berlin-Heidelberg-New York 1966.

[Ko] H.J.Kowalsky: Limesräume und Komplettierung. Math.Nachr. 12 (1954), 301-340.

[Ku 1] K.Kutzler: Eine Charakterisierung von Lindelöfräumen. To appear in Arch.Math.

[Ku 2] ——: Über Zusammenhänge, die zwischen einigen Limitierungen auf C(X) und dem Satz von Dini bestehen. Habilitationsschrift at the University of Mannheim (1972).

——: Bemerkungen über unendlichdimensionale, separierte Limesvektorräume und Limesgruppen. J.Reine u.Angew.Math. 253 (1972), 98-116.

——: Über einige Limitierungen auf C(X) und den Satz von Dini. To appear in Math.Nachr.

[Ku 3] ——: Eine Bemerkung über endlichdimensionale, separierte, limitierte Vektorräume. Arch.Math. XX, Fasc.2 (1969), 165-168.

[Ma] G.Marinescu: Espaces vectoriels pseudotopologiques et théorie des distributions. VEB Deutscher Verlag der Wissenschaften 1963.

J.A.Leslie: On differential structure for the group of diffeomorphisms. Topology, Bd. 6, 263-271 (1967).

[Mo,Wu] P.D.Morris, D.E.Wulbert: Functional representation of topological algebras. Pac.J.Math.Vol.22, No.2 (1967), 323-337.

[Mü] B.Müller: Über den c-Dual eines Limesvektorraumes. Ph.D.Thesis, University of Mannheim (1972).

——: Über die Charakterisierung c-einbettbarer, topologischer Räume X durch $C_{co}(X)$. To appear in Arch. Math.

——: Dualitätstheorie für Vektorunterverbände von $C_c(X)$. To appear in Mathematische Nachrichten.

——: L_c- und c-einbettbare Limesräume. To appear in Ann. Scuola Normale Sup., Pisa.

[Na] L.Nachbin: Elements of approximation theory. Van Nostrand Math.Studies 14. Van Nostrand, Princeton, 1967.

——: Topological vector spaces of continuous functions. Proc.Nat.Acad.Sci.USA 40 (1954), 471-474.

[Po] H.Poppe: Compactness in general function spaces. VEB Deutscher Verlag der Wissenschaften, Berlin (1974).

[Ra,Wy] J.F.Ramaley and O.Wyler: Cauchy-spaces II: Regular completions and Compactifications. Math.Ann.187, p. 187-199.

——: J.F.Ramaley and O.Wyler: Cauchy spaces I: Structure and Uniformization. Theorems, Math.Ann. 197, p.175-186.

[Ri] C.E.Rickart: Banach algebras. Van Nostrand Princeton 1960.

[Schae] H.H.Schaefer: Stetige Konvergenz in allgemeinen topologischen Räumen. Arch.Math.6 (1955), 423-427.

[Sch 1] M.Schroder: Continuous convergence in a Gelfand theory for topological algebras. Ph.D.Thesis,Queen's Univ., Kingston, Canada (1971).

[Sch 2] ——: Characterizations of c-embedded spaces. Preprint.

——: Solid convergence spaces. Bull.Austr.Math.Soc.Vol.8 (1973), 443-459.

——: A family of c-embedded spaces whose associated completely regular topology is compact. Arch.Math.Vol.XXV, (1974), 69-74.

——: A characterization of c-embedded convergence Spaces, Research Report, No.18 (1972), University of Waikato, Hamilton, New Zealand.

——: The Structures of μ-convergence. Research Report, No. 22 (1974), University of Waikato, Hamilton, New Zealand.

——: Notes on the c-duality of convergence vector spaces. Queen's Mathematical Preprints 35, Kingston 1971.

Z.Semadeni: Banach spaces of continuous functions. Vol.1, Scient.Publishers Warszawa (1971).

T.Shirota: On locally convex vector spaces of continuous functions. Proc.Japan Acad.30 (1954), 294-298.

[Wl] J.Wloka: Limesräume und Distributionen. Math.Ann.152 (1963),351-409.

[Wa] S.Warner: The topology of compact convergence on continuous function spaces. Duke Math. J.25 (1958), 265-282.

[Wo] M.Wolff: Nonstandard Komplettierung von Cauchy-Algebren, in: Contributions to Nonstandard Analysis ed. by W.Luxemburg and A. Robinson (p.179-213).North Holland Pub.Comp. Amsterdam - London 1972.

I N D E X

LIST OF SOME SYMBOLS

Sets with the indices c, co and s carry the continuous convergence structure, the topology of compact convergence and the topology of pointwise convergence, respectively.

Vol. 309: D. H. Sattinger, Topics in Stability and Bifurcation Theory. VI, 190 pages. 1973. DM 20,–

Vol. 310: B. Iversen, Generic Local Structure of the Morphisms in Commutative Algebra. IV, 108 pages. 1973. DM 18,–

Vol. 311: Conference on Commutative Algebra. Edited by J. W. Brewer and E. A. Rutter. VII, 251 pages. 1973. DM 24,–

Vol. 312: Symposium on Ordinary Differential Equations. Edited by W. A. Harris, Jr. and Y. Sibuya. VIII, 204 pages. 1973. DM 22,–

Vol. 313: K. Jörgens and J. Weidmann, Spectral Properties of Hamiltonian Operators. III, 140 pages. 1973. DM 18,–

Vol. 314: M. Deuring, Lectures on the Theory of Algebraic Functions of One Variable. VI, 151 pages. 1973. DM 18,–

Vol. 315: K. Bichteler, Integration Theory (with Special Attention to Vector Measures). VI, 357 pages. 1973. DM 29,–

Vol. 316: Symposium on Non-Well-Posed Problems and Logarithmic Convexity. Edited by R. J. Knops. V, 176 pages. 1973. DM 20,–

Vol. 317: Séminaire Bourbaki – vol. 1971/72. Exposés 400–417. IV, 361 pages. 1973. DM 29,–

Vol. 318: Recent Advances in Topological Dynamics. Edited by A. Beck. VIII, 285 pages. 1973. DM 27,–

Vol. 319: Conference on Group Theory. Edited by R. W. Gatterdam and K. W. Weston. V, 188 pages. 1973. DM 20,–

Vol. 320: Modular Functions of One Variable I. Edited by W. Kuyk. V, 195 pages. 1973. DM 20,–

Vol. 321: Séminaire de Probabilités VII. Edité par P. A. Meyer. VI, 322 pages. 1973. DM 29,–

Vol. 322: Nonlinear Problems in the Physical Sciences and Biology. Edited by I. Stakgold, D. D. Joseph and D. H. Sattinger. VIII, 357 pages. 1973. DM 29,–

Vol. 323: J. L. Lions, Perturbations Singulières dans les Problèmes aux Limites et en Contrôle Optimal. XII, 645 pages. 1973. DM 46,–

Vol. 324: K. Kreith, Oscillation Theory. VI, 109 pages. 1973. DM 18,–

Vol. 325: C.-C. Chou, La Transformation de Fourier Complexe et L'Equation de Convolution. IX, 137 pages. 1973. DM 18,–

Vol. 326: A. Robert, Elliptic Curves. VIII, 264 pages. 1973. DM 24,–

Vol. 327: E. Matlis, One-Dimensional Cohen-Macaulay Rings. XII, 157 pages. 1973. DM 20,–

Vol. 328: J. R. Büchi and D. Siefkes, The Monadic Second Order Theory of All Countable Ordinals. VI, 217 pages. 1973. DM 22,–

Vol. 329: W. Trebels, Multipliers for (C, α)-Bounded Fourier Expansions in Banach Spaces and Approximation Theory. VII, 103 pages. 1973. DM 18,–

Vol. 330: Proceedings of the Second Japan-USSR Symposium on Probability Theory. Edited by G. Maruyama and Yu. V. Prokhorov. VI, 550 pages. 1973. DM 40,–

Vol. 331: Summer School on Topological Vector Spaces. Edited by L. Waelbroeck. VI, 226 pages. 1973. DM 22,–

Vol. 332: Séminaire Pierre Lelong (Analyse) Année 1971-1972. V, 131 pages. 1973. DM 18,–

Vol. 333: Numerische, insbesondere approximationstheoretische Behandlung von Funktionalgleichungen. Herausgegeben von R. Ansorge und W. Törnig. VI, 296 Seiten. 1973. DM 27,–

Vol. 334: F. Schweiger, The Metrical Theory of Jacobi-Perron Algorithm. V, 111 pages. 1973. DM 18,–

Vol. 335: H. Huck, R. Roitzsch, U. Simon, W. Vortisch, R. Walden, B. Wegner und W. Wendland, Beweismethoden der Differentialgeometrie im Großen. IX, 159 Seiten. 1973. DM 20,–

Vol. 336: L'Analyse Harmonique dans le Domaine Complexe. Edité par E. J. Akutowicz. VIII, 169 pages. 1973. DM 20,–

Vol. 337: Cambridge Summer School in Mathematical Logic. Edited by A. R. D. Mathias and H. Rogers. IX, 660 pages. 1973. DM 46,–

Vol: 338: J. Lindenstrauss and L. Tzafriri, Classical Banach Spaces. IX, 243 pages. 1973. DM 24,–

Vol. 339: G. Kempf, F. Knudsen, D. Mumford and B. Saint-Donat, Toroidal Embeddings I. VIII, 209 pages. 1973. DM 22,–

Vol. 340: Groupes de Monodromie en Géométrie Algébrique. (SGA 7 II). Par P. Deligne et N. Katz. X, 438 pages. 1973. DM 44,–

Vol. 341: Algebraic K-Theory I, Higher K-Theories. Edited by H. Bass. XV, 335 pages. 1973. DM 29,–

Vol. 342: Algebraic K-Theory II, "Classical" Algebraic K-Theory, and Connections with Arithmetic. Edited by H. Bass. XV, 527 pages. 1973. DM 40,–

Vol. 343: Algebraic K-Theory III, Hermitian K-Theory and Geometric Applications. Edited by H. Bass. XV, 572 pages. 1973. DM 40,–

Vol. 344: A. S. Troelstra (Editor), Metamathematical Investigation of Intuitionistic Arithmetic and Analysis. XVII, 485 pages. 1973. DM 38,–

Vol. 345: Proceedings of a Conference on Operator Theory. Edited by P. A. Fillmore. VI, 228 pages. 1973. DM 22,–

Vol. 346: Fučík et al., Spectral Analysis of Nonlinear Operators. II, 287 pages. 1973. DM 26,–

Vol. 347: J. M. Boardman and R. M. Vogt, Homotopy Invariant Algebraic Structures on Topological Spaces. X, 257 pages. 1973. DM 24,–

Vol. 348: A. M. Mathai and R. K. Saxena, Generalized Hypergeometric Functions with Applications in Statistics and Physical Sciences. VII, 314 pages. 1973. DM 26,–

Vol. 349: Modular Functions of One Variable II. Edited by W. Kuyk and P. Deligne. V, 598 pages. 1973. DM 38,–

Vol. 350: Modular Functions of One Variable III. Edited by W. Kuyk and J.-P. Serre. V, 350 pages. 1973. DM 26,–

Vol. 351: H. Tachikawa, Quasi-Frobenius Rings and Generalizations. XI, 172 pages. 1973. DM 20,–

Vol. 352: J. D. Fay, Theta Functions on Riemann Surfaces. V, 137 pages. 1973. DM 18,–

Voi. 353: Proceedings of the Conference on Orders, Group Rings and Related Topics. Organized by J. S. Hsia, M. L. Madan and T. G. Ralley. X, 224 pages. 1973. DM 22,–

Vol. 354: K. J. Devlin, Aspects of Constructibility. XII, 240 pages. 1973. DM 24,–

Vol. 355: M. Sion, A Theory of Semigroup Valued Measures. V, 140 pages. 1973. DM 18,–

Vol. 356: W. L. J. van der Kallen, Infinitesimally Central-Extensions of Chevalley Groups. VII, 147 pages. 1973. DM 18,–

Vol. 357: W. Borho, P. Gabriel und R. Rentschler, Primideale in Einhüllenden auflösbarer Lie-Algebren. V, 182 Seiten. 1973. DM 20,–

Vol. 358: F. L. Williams, Tensor Products of Principal Series Representations. VI, 132 pages. 1973. DM 18,–

Vol. 359: U. Stammbach, Homology in Group Theory. VIII, 183 pages. 1973. DM 20,–

Vol. 360: W. J. Padgett and R. L. Taylor, Laws of Large Numbers for Normed Linear Spaces and Certain Fréchet Spaces. VI, 111 pages. 1973. DM 18,–

Vol. 361: J. W. Schutz, Foundations of Special Relativity: Kinematic Axioms for Minkowski Space Time. XX, 314 pages. 1973. DM 26,–

Vol. 362: Proceedings of the Conference on Numerical Solution of Ordinary Differential Equations. Edited by D. Bettis. VIII, 490 pages. 1974. DM 34,–

Vol. 363: Conference on the Numerical Solution of Differential Equations. Edited by G. A. Watson. IX, 221 pages. 1974. DM 20,–

Vol. 364: Proceedings on Infinite Dimensional Holomorphy. Edited by T. L. Hayden and T. J. Suffridge. VII, 212 pages. 1974. DM 20,–

Vol. 365: R. P. Gilbert, Constructive Methods for Elliptic Equations. VII, 397 pages. 1974. DM 26,–

Vol. 366: R. Steinberg, Conjugacy Classes in Algebraic Groups (Notes by V. V. Deodhar). VI, 159 pages. 1974. DM 18,–

Vol. 367: K. Langmann und W. Lütkebohmert, Cousinverteilungen und Fortsetzungssätze. VI, 151 Seiten. 1974. DM 16,–

Vol. 368: R. J. Milgram, Unstable Homotopy from the Stable Point of View. V, 109 pages. 1974. DM 16,–

Vol. 369: Victoria Symposium on Nonstandard Analysis. Edited by A. Hurd and P. Loeb. XVIII, 339 pages. 1974. DM 26,–

Vol. 370: B. Mazur and W. Messing, Universal Extensions and One Dimensional Crystalline Cohomology. VII, 134 pages. 1974. DM 16,–

Vol. 371: V. Poenaru, Analyse Différentielle. V, 228 pages. 1974. DM 20,–

Vol. 372: Proceedings of the Second International Conference on the Theory of Groups 1973. Edited by M. F. Newman. VII, 740 pages. 1974. DM 48,–

Vol. 373: A. E. R. Woodcock and T. Poston, A Geometrical Study of the Elementary Catastrophes. V, 257 pages. 1974. DM 22,–

Vol. 374: S. Yamamuro, Differential Calculus in Topological Linear Spaces. IV, 179 pages. 1974. DM 18,–

Vol. 375: Topology Conference 1973. Edited by R. F. Dickman Jr. and P. Fletcher. X, 283 pages. 1974. DM 24,–

Vol. 376: D. B. Osteyee and I. J. Good, Information, Weight of Evidence, the Singularity between Probability Measures and Signal Detection. XI, 156 pages. 1974. DM 16.–

Vol. 377: A. M. Fink, Almost Periodic Differential Equations. VIII, 336 pages. 1974. DM 26,–

Vol. 378: TOPO 72 – General Topology and its Applications. Proceedings 1972. Edited by R. Alò, R. W. Heath and J. Nagata. XIV, 651 pages. 1974. DM 50,–

Vol. 379: A. Badrikian et S. Chevet, Mesures Cylindriques, Espaces de Wiener et Fonctions Aléatoires Gaussiennes. X, 383 pages. 1974. DM 32,–

Vol. 380: M. Petrich, Rings and Semigroups. VIII, 182 pages. 1974. DM 18,–

Vol. 381: Séminaire de Probabilités VIII. Edité par P. A. Meyer. IX, 354 pages. 1974. DM 32,–

Vol. 382: J. H. van Lint, Combinatorial Theory Seminar Eindhoven University of Technology. VI, 131 pages. 1974. DM 18,–

Vol. 383: Séminaire Bourbaki – vol. 1972/73. Exposés 418-435 IV, 334 pages. 1974. DM 30,–

Vol. 384: Functional Analysis and Applications, Proceedings 1972. Edited by L. Nachbin. V, 270 pages. 1974. DM 22,–

Vol. 385: J. Douglas Jr. and T. Dupont, Collocation Methods for Parabolic Equations in a Single Space Variable (Based on C^1-Piecewise-Polynomial Spaces). V, 147 pages. 1974. DM 16,–

Vol. 386: J. Tits, Buildings of Spherical Type and Finite BN-Pairs. IX, 299 pages. 1974. DM 24,–

Vol. 387: C. P. Bruter, Eléments de la Théorie des Matroïdes. V, 138 pages. 1974. DM 18,–

Vol. 388: R. L. Lipsman, Group Representations. X, 166 pages. 1974. DM 20,–

Vol. 389: M.-A. Knus et M. Ojanguren, Théorie de la Descente et Algèbres d' Azumaya. IV, 163 pages. 1974. DM 20,–

Vol. 390: P. A. Meyer, P. Priouret et F. Spitzer, Ecole d'Eté de Probabilités de Saint-Flour III – 1973. Edité par A. Badrikian et P.-L. Hennequin. VIII, 189 pages. 1974. DM 20,–

Vol. 391: J. Gray, Formal Category Theory: Adjointness for 2-Categories. XII, 282 pages. 1974. DM 24,–

Vol. 392: Géométrie Différentielle, Colloque, Santiago de Compostela, Espagne 1972. Edité par E. Vidal. VI, 225 pages. 1974. DM 20,–

Vol. 393: G. Wassermann, Stability of Unfoldings. IX, 164 pages. 1974. DM 20,–

Vol. 394: W. M. Patterson 3rd, Iterative Methods for the Solution of a Linear Operator Equation in Hilbert Space – A Survey. III, 183 pages. 1974. DM 20,–

Vol. 395: Numerische Behandlung nichtlinearer Integrodifferential- und Differentialgleichungen. Tagung 1973. Herausgegeben von R. Ansorge und W. Törnig. VII, 313 Seiten. 1974. DM 28,–

Vol. 396: K. H. Hofmann, M. Mislove and A. Stralka, The Pontryagin Duality of Compact O-Dimensional Semilattices and its Applications. XVI, 122 pages. 1974. DM 18,–

Vol. 397: T. Yamada, The Schur Subgroup of the Brauer Group. V, 159 pages. 1974. DM 18,–

Vol. 398: Théories de l'Information, Actes des Rencontres de Marseille-Luminy, 1973. Edité par J. Kampé de Fériet et C. Picard. XII, 201 pages. 1974. DM 23,–

8480-43
5-20

Vol. 399: Functional Analysis and its Applications, Proceedings 1973. Edited by H. G. Garnir, K. R. Unni and J. H. Williamson. XVII, 569 pages. 1974. DM 44,–

Vol. 400: A Crash Course on Kleinian Groups – San Francisco 1974. Edited by L. Bers and I. Kra. VII, 130 pages. 1974. DM 18,–

Vol. 401: F. Atiyah, Elliptic Operators and Compact Groups. V, 93 pages. 1974. DM 18,–

Vol. 402: M. Waldschmidt, Nombres Transcendants. VIII, 277 pages. 1974. DM 25,–

Vol. 403: Combinatorial Mathematics – Proceedings 1972. Edited by D. A. Holton. VIII, 148 pages. 1974. DM 18,–

Vol. 404: Théorie du Potentiel et Analyse Harmonique. Edité par J. Faraut. V, 245 pages. 1974. DM 25,–

Vol. 405: K. Devlin and H. Johnsbråten, The Souslin Problem. VIII, 132 pages. 1974. DM 18,–

Vol. 406: Graphs and Combinatorics – Proceedings 1973. Edited by R. A. Bari and F. Harary. VIII, 355 pages. 1974. DM 30,–

Vol. 407: P. Berthelot, Cohomologie Cristalline des Schémas de Caracteristique p > o. VIII, 598 pages. 1974. DM 44,–

Vol. 408: J. Wermer, Potential Theory. VIII, 146 pages. 1974. DM 18,–

Vol. 409: Fonctions de Plusieurs Variables Complexes, Séminaire François Norguet 1970–1973. XIII, 612 pages. 1974. DM 47,–

Vol. 410: Séminaire Pierre Lelong (Analyse) Année 1972–1973. VI, 181 pages. 1974. DM 18,–

Vol. 411: Hypergraph Seminar. Ohio State University, 1972. Edited by C. Berge and D. Ray-Chaudhuri. IX, 287 pages. 1974. DM 28,–

Vol. 412: Classification of Algebraic Varieties and Compact Complex Manifolds. Proceedings 1974. Edited by H. Popp. V, 333 pages. 1974. DM 30,–

Vol. 413: M. Bruneau, Variation Totale d'une Fonction. XIV, 332 pages. 1974. DM 30,–

Vol. 414: T. Kambayashi, M. Miyanishi and M. Takeuchi, Unipotent Algebraic Groups. VI, 165 pages. 1974. DM 20,–

Vol. 415: Ordinary and Partial Differential Equations, Proceedings of the Conference held at Dundee, 1974. XVII, 447 pages. 1974. DM 37,–

Vol. 416: M. E. Taylor, Pseudo Differential Operators. IV, 155 pages. 1974. DM 18,–

Vol. 417: H. H. Keller, Differential Calculus in Locally Convex Spaces. XVI, 131 pages. 1974. DM 18,–

Vol. 418: Localization in Group Theory and Homotopy Theory and Related Topics Battelle Seattle 1974 Seminar. Edited by P. J. Hilton. VI, 171 pages. 1974. DM 20,–

Vol. 419: Topics in Analysis – Proceedings 1970. Edited by O. E. Lehto, I. S. Louhivaara, and R. H. Nevanlinna. XIII, 391 pages. 1974. DM 35,–

Vol. 420: Category Seminar. Proceedings, Sydney Category Theory Seminar 1972/73. Edited by G. M. Kelly. VI, 375 pages. 1974. DM 32,–

Vol. 421: V. Poénaru, Groupes Discrets. VI, 216 pages. 1974. DM 23,–

Vol. 422: J.-M. Lemaire, Algèbres Connexes et Homologie des Espaces de Lacets. XIV, 133 pages. 1974. DM 23,–

Vol. 423: S. S. Abhyankar and A. M. Sathaye, Geometric Theory of Algebraic Space Curves. XIV, 302 pages. 1974. DM 28,–

Vol. 424: L. Weiss and J. Wolfowitz, Maximum Probability Estimators and Related Topics. V, 106 pages. 1974. DM 18,–

Vol. 425: P. R. Chernoff and J. E. Marsden, Properties of Infinite Dimensional Hamiltonian Systems. IV, 160 pages. 1974. DM 20,–

Vol. 426: M. L. Silverstein, Symmetric Markov Processes. IX, 287 pages. 1974. DM 28,–

Vol. 427: H. Omori, Infinite Dimensional Lie Transformation Groups. XII, 149 pages. 1974. DM 18,–

Vol. 428: Algebraic and Geometrical Methods in Topology, Proceedings 1973. Edited by L. F. McAuley. XI, 280 pages. 1974. DM 28,–